P9-DWK-446

Where Land and Sea Meet

The Earth, Its Wonders, Its Secrets

Where Land and Sea Meet

THE READER'S DIGEST ASSOCIATION, INC.
Pleasantville, New York/Montreal

WHERE LAND AND SEA MEET
Edited and designed by Toucan Books Limited
with Bradbury and Williams
Written by John Man
Edited by Helen Douglas-Cooper
Picture Research by Marian Pullen

FOR THE READER'S DIGEST UK
Series Editor: Christine Noble
Editorial Assistant: Chloe Garrow
Editorial Director: Cortina Butler
Art Director: Nick Clark

READER'S DIGEST, US
Senior Editor: Fred Dubose
Senior Designer: Judith Carmel
Group Editorial Director, Nature: Wayne Kalyn
Vice President, Editor-in-Chief: Christopher Cavanaugh
Art Director: Joan Mazzeo

Reprinted with amendments 1999

Separations: David Bruce Graphics Limited, London

Printed in the United States of America, 1999

Library of Congress Cataloging in Publication Data

Where land and sea meet.
p. cm. -- (The earth, its wonders, its secrets)
Includes index.
ISBN 0-7621-0114-8 (hardcover)
1. Coasts. 2. Coastal ecology. 3. Islands.
I. Reader's Digest Association. II. Series.
GB451.2 .W48 1999
551.45'7 – dc21

98-25946

FRONT COVER *Pebbles from the beach of Skeleton Coast National Park, Namibia. Inset: Australian sea lions* (Neophoca cinerea) *bask on the beach of Kangaroo Island, Australia.*

PAGE 3 *A wave-cut platform of carboniferous limestone forms the cliffs at St Govan's Head, Pembrokeshire, Wales.*

Contents

THE EVER-CHANGING COAST

The constant battle between land and sea has carved out an endless variety of coastlines – from wild surf beaches to steep, rocky inlets, and from salt marshes to coral reefs – that are among the richest and most productive habitats on Earth.

The coast – the margin where land meets sea – is an exceptional place providing living space for an extraordinary diversity of plants and animals. There are rocky shores backed by high, precipitous cliffs, such as those of Cornwall, Ireland and Brittany, where seals haul out and hide in sea caves, sea birds nest on cliff-side ledges, and sea anemones and tiny shrimp-like creatures, stranded in shallow rock pools by the receding tide, fall prey to both sea birds and land birds; there are expansive beaches, such as those of the coast of northern France and the Bay of Biscay, where flatfish swim close to shore and burrowing bivalve molluscs take cover beneath the sand; and there are estuaries and mud flats, such as Bridgwater Bay in Somerset, England, and the Dee Estuary on the east coast of Scotland, where thousands of waders come to feed on the life – worms, crustaceans and molluscs – hidden below the surface.

The coast is the place where the plants and animals of one world meet those of another; it is where salt water meets fresh, heat meets cold, dry meets wet, and where living things must adapt to the regularly changing conditions if they are to survive. It is often a turbulent place, harsh and inhospitable, a battleground between land, sea and air; it is a transient zone, where the ground can appear or disappear overnight; and it is often a brutal place where powerful waves, having travelled as gentle undulations across hundreds of miles of ocean, gathering energy as they go, slam into the shoreline with a force of several tons per square foot, breaking apart rocks and destroying cliffs and coastland.

Even on a calm day, an average of five waves break against a section of shore every minute, and between waves the backwash drags rocks and debris back out to sea. The churning water grinds the fragments against one another to form shingle and sand, and then throws them up against the shore once more to form beaches, bars and spits. Once or twice in each 24 hours, the intertidal zone is drowned by the incoming tide, and exposed to the air and the elements when it goes out again. Rocks, pebbles and sand, with the plants and animals that manage to find a way of living among them, may be battered by wind, drenched by rain or scorched by the sun when the tide is out, and swamped by salt water and pounded by waves during the daily incursions of the sea.

In such shifting surroundings and extreme conditions, it is remarkable that life exists at all. Yet, despite this seeming inhospitality, there is a profusion of life on the coast and in coastal waters. All the requirements for life – light, water, minerals and oxygen – are present in abundance, and living things, always alert for opportunities to colonise new and vacant habitats, have invaded the shore and relatively shallow inshore waters.

These colonists have developed adaptations that enable them to survive in such

STICKY ARM *A common sun star has a remarkable system of tube feet on the underside of its arms, which anchor it to rocks.*

WAVE-BATTERED COAST *The Dingle Peninsula in western Ireland takes the full brunt of waves that have formed in the Atlantic Ocean.*

Plants Take Over *In quiet backwaters mangroves help the land to invade the sea by anchoring sand, mud and soil in place.*

unstable environments. Some root themselves firmly in place: seaweeds have holdfasts (tough root-like structures at their base); barnacles use a powerful protein cement; mussels are fastened by strong protein threads; and starfish have hundreds of tiny sucker-tipped 'tube feet' to anchor themselves to rocks pounded constantly by the waves. Others wedge themselves into cracks and crevices, or are lost in a tangle of mangroves or seaweeds. On muddy shores, animals such as worms, clams and ghost shrimps avoid the twice-daily exposure to the air by hiding in burrows beneath the beach or mud flats, while others, such as shore crabs, cower in cavities beneath large boulders.

Primeval Forces *Glowing lava and smoke spew from the volcano on Surtsey, creating new land off the southern coast of Iceland.*

Constant Change

While the sea invades the land in one place, along another stretch of coastline the land encroaches upon the sea. On sheltered shorelines, away from the destructive power of the waves, salt-resistant plants such as cordgrasses and mangroves spearhead an advance that inches the land forwards year by year. Also, sands are shifted by the currents and waves, sealing off arms of the sea. The resulting shallow lagoons, invaded by plants from the shore, gradually fill with silt and decaying organic materials to become land.

Some legends have it that the land was originally born out of the sea. Several North American Indian tribes, for example, shared a similar story of the creation of the continents. They believed that the Earth was once a vast and featureless ocean, and that the only living thing was a huge bird. A glance from its fiery eyes caused flashes of lightning to crisscross the skies, and the beat of its immense wings made a roar like thunder. Flying down to the great ocean, it touched the water and the land rose up through the waves.

It is easy to see how such ancient myths might have arisen, as was witnessed on the morning of November 14, 1963, when a dramatic event started to shake the North Atlantic, near the Westman Islands, off southern Iceland. The surface of the sea began to boil, and beneath the turbulence the water glowed red, while billowing clouds of ash and large, fiery rocks were hurled into the air. By the morning of the following day, a glowing cloud of cinders pushed above the waves and gradually grew as 400 000 tons of ash and volcanic bombs (lumps of solidified lava) were thrown out every hour. The ash and lava fragments formed into a long ridge, with several fissures erupting simultaneously. The sea poured into the vents and was promptly blown back out again.

By February 1964 the newborn island – christened Surtsey after the Norse god Sutur, who brought fire from the south and fought against Freya, the goddess of fertility – was just over 1/3 sq mile (1 km^2) in area and had risen 570 ft (174 m) above sea level. The sea, though, pounded the crumbly shores, and rain and wind etched deep rivulets in the ash slopes. In the fight between the elements, the land seemed to be losing; the atmosphere and the ocean were slicing it away.

Then, in April, a new crater emerged. Named Sutur II, it disgorged a continuous stream of lava at an incredible rate. Almost 20 tons per second, at a temperature of 1100°C (2012°F), rushed down the slopes at a speed of about 45 mph (70 km/h). When it hit the water, clouds of steam rose from the sea, and the lava solidified into a hard, protective skin. The land was on the offensive again, and by June 1964 the island had grown to just over 1 sq mile (3 km^2). It was time for colonisation.

Surtsey became a natural outdoor laboratory where scientists could observe every

VIOLENT ERUPTION *Surtsey explodes from the surface of the sea in a series of eruptions that eventually produced enough lava and ash to resist the action of the waves. Its volcanic slopes were barren at first (left), but life was quick to arrive.*

aspect of island formation and the mechanisms of colonisation of a brand new stretch of coast from their very beginnings. The first living organisms to appear were bacteria, although a larger visitor was a fly (*Diamesa zernyi*). A variety of sea birds – kittiwakes (*Rissa tridactyla*), great black-backed gulls (*Larus marinus*) and glaucous

Ancient Art *This mosaic of a sea creature, dating from the 3rd or 2nd century BC, comes from Delos in Greece.*

gulls (*Larus hyperboreus*), as well as a solitary Arctic tern (*Sterna paradisaea*) – landed on the beaches between eruptions. On June 3, 1965, the first plants were recorded – 40 seedlings of sea rocket (*Cakile maritima*), a straggly stemmed plant that inhabits the higher levels of European beaches and whose fruits are dispersed by tides and sea currents.

By 1967, 30 bird species had visited the island; 14 species of seaweeds grew along its edges; colonies of the moss *Funaria hygrometrica* and two species of lichens were established on the craters and cliffs, while seals hauled themselves out onto the black, glassy-sand beaches. Two years later, 29 species of migratory birds had flown down to investigate the new land, and by 1971 the first fulmars (*Fulmaris glacialis*) and black guillemots (*Cepphus grylle*) nested in cracks and crevices in the lava rocks.

The Continental Shelf

The extravagance of life found along coasts and in coastal waters is the result of the way in which land and sea overlap. Around any coast is a strip of seabed, covered by shallow seas up to 660 ft (200 m) deep and extending out to the edge of underwater slopes and cliffs that fall away into the ocean's abyss. Known as the continental shelf, this area is actually a gradually sloping extension of the land, although it is covered by water today. In different parts of the world it varies considerably in width. Around the margins of the Atlantic Ocean, for example, the broadest continental shelf is that of north-western Europe, on which sits the United Kingdom and Ireland, plus the North Sea, the Skagerrak between Denmark and Norway, and the English Channel. The narrowest part is off Cape Canaveral on the east coast of North America, where the shelf is almost non-existent.

Sea levels rise and fall over time as the ice in polar regions forms or melts, and the continental shelf, which constitutes about one-tenth of the seabed, has been drowned during some periods of the Earth's history and exposed at others. When covered by water, this narrow strip is probably the most productive part of the entire marine environment, supporting some of the richest and most varied habitats on Earth, and rivalling the tropical rain forests in the diversity and quantity of wildlife.

In some places, upwellings (cold water that rises from the depths of the sea) carry nutrients from the seabed up to the surface, providing food for drifting microscopic plants (the phytoplankton). Elsewhere, light from the Sun can penetrate the relatively shallow waters over the continental shelf to the seabed, providing enough energy at certain times of the year for phytoplankton to flourish. The phytoplankton provide food for microscopic animals (the zooplankton), which are then eaten by small fish; these in turn are eaten by larger fish and marine animals, and so on progressively up the food chain. With such an abundance of food in temperate latitudes during spring and summer, the continental shelf waters are filled with fish, squid and other marine life. Some, like the gigantic basking shark (*Cetorhinus maximus*), come to feed, while others, such as herring, arrive in inshore waters in their millions to spawn. Their eggs and milt (sperm) turn coastal waters milky, and their offspring feed on the supply of microscopic food.

Animals, both marine and terrestrial, take advantage of the abundance, their paths often crossing. Sea birds, such as gulls and auks, nest on cliffs, but rely mainly on the shore and shallow seas for food. When they find a shoal of herring or sand eels, the

Sea Harvest *Madagascan children beat the water with twigs to drive fish into their catching-basket. Below: A Norwegian fishing boat negotiates a rocky coastline as it returns to port with its catch.*

birds flock to the sea's surface, the ensuing commotion providing a beacon for pods of minke whales (*Balaenoptera acutorostrata*) – small baleen whales that patrol continental shelves worldwide – which have learned to associate even a fleeting concentration of birds with a readily available meal.

On the outer fringes of continental shelves, islands and shallow banks support coral reefs, which may be so extensive that they build into continuous coral barriers, like the Great Barrier Reef off Queensland, Australia, the largest living entity on Earth. These underwater jewelled gardens are so filled with living things that scientists are discovering new species every day. During the 1970s, for example, 40 per cent of fish examined by researchers exploring the coral reefs of the Bahamas, in the western Atlantic, were new to science. It is thought by scientists that three-quarters of all the fish in the sea – and most are found over the continental shelves – have yet to be discovered.

With such a wealth of resources it was inevitable that coastal areas would become inhabited by people too. Thousands of years ago, humankind discovered the benefits of coastal living, taking advantage of the abundance of food and milder climates to found the first coastal settlements.

The Perils of Pollution

Inshore waters may provide nurseries, feeding sites and living space for the greater portion of life in the sea, as well as for some land-based creatures, but they are also used as a 'dustbin' for human waste. Birds of prey that feed around the shores of north-western Europe have been poisoned by pesticides; the white whales (*Delphinapterus leucas*) of the St Lawrence River on Canada's east coast are unable to breed because of chemical waste; polar bears (*Ursus maritimus*) wandering the icy wastes of the Arctic are so filled with pollutants that they could be officially declared 'hazardous waste'; and each summer a 230 mile (375 km) stretch of Italy's Adriatic coast is covered by a yellowish-brown slime, the result of high levels of fertilisers from agriculture.

Oil-spill disasters hit the headlines when thousands of sea birds and marine creatures are smothered and killed by thick, suffocating oil. However, some less

WASTE AND MORE WASTE *A sewage pipeline is surrounded by milky-coloured industrial effluent, while raw sewage pumped into the sea (left) contains waste products that could be harmful to wildlife.*

OIL-SPILL VICTIM *Despite being cleaned up, victims of oil spills usually die soon after release, probably as a result of stress. Right: An oil slick washed up on a Texas beach coats sand and pebbles.*

well-known pollution events, like the spill of 100 000 barrels (3.5 million gallons) of crude oil on the north coast of Panama in 1986, are thought to harbour less obvious but longer-term dangers for marine life. The slick spread along 50 miles (80 km) of coral reef and mangrove forest, killing the plants and animals living there. This coral and mangrove barrier, which up to then had cushioned the coast from the onslaught of the waves, was breached, and serious erosion followed. Coral was killed down to a depth of 20 ft (6 m), sea grasses were eliminated entirely, a third of the mangroves died, and grazing fish, which kept the reef clear of blanketing algae, were so reduced in numbers that the remaining portions of the reef are now being smothered by a carpet of green slime. Every time it rains, trapped oil oozes out to form new slicks, and scientists believe it will take over a century for the reef to recover.

A new, more insidious form of pollution, only recently recognised as a serious threat to life in coastal waters, is caused by the transport of organisms from one part of the globe to another. Ships carry not only land animals, such as rats and cats, which jump ship and plague remote islands, but also marine organisms. Ships' ballast is like a mobile aquarium in which stowaways can travel the seven seas. When bilges are cleaned and the effluent discharged, often in harbour, all manner of life-forms are off-loaded too, with potentially long-lasting and devastating ecological effects.

The organisms responsible for cholera and botulism, and the dinoflagellates that cause toxic blooms, have been found in ships' ballast tanks. A starfish (*Asterias amurensis*)

Red Tide *A bloom of toxic dinoflagellates threatens marine creatures at Whangarei Heads on New Zealand's North Island.*

found in the northern Pacific, and most prevalent on the Japanese coast, has found its way to Tasmania, where it attacks shellfish and other marine life; the giant seaweed *Undaria pinnatifida* has overrun parts of the New Zealand coast; and in the Mediterranean the tropical green alga *Caulerpa tarjolia* has ousted sea grasses vital to young fish.

Despite international treaties aimed at conserving inshore waters, the trend is proving slow to reverse, although there are glimmers of hope that humankind is beginning to realise the fragility of natural coastal communities. We are, like our ancient forebears, excited by the wonder of life along the coast: whale-catching has been replaced by whale-watching; scuba-diving and snorkelling have become popular holiday pastimes; planning authorities in some parts of the world, such as Costa Rica, are insisting that ecotourism relies on facilities that blend in with the environment; and artificial reefs, made of anything from car tyres to old barges, have been sunk off coasts to create new habitats for fish and other marine life.

The quest to unravel the complexities of coastal ecosystems, in order to understand the interrelationships of plants and animals within them, continues. In the Gulf of Maine, on the east coast of North America, for example, scientists from Northeastern University, Boston, have discovered the impact of cod fishing on the relationship between cod, sea urchins and the plants that live on the seabed. Cod are major predators of sea urchins, and where they have been fished intensively the sea urchins thrive. Huge, marauding crowds of sea urchins, each group over 1000 strong, devour every living brown seaweed in their path, reducing vast tracts of seabed from Massachusetts to Nova Scotia to barren deserts. Where cod keep the sea urchins under control, the sea floor is covered by a healthy forest of kelp which is home to all manner of marine life, including the young of many species of fish.

Giant Seaweed *A giant Japanese seaweed accidentally introduced to the British south coast is swamping the indigenous seaweeds.*

A Delicate Skin

Because the coast is a place where land-based animals go to feed and where marine life, including open-ocean interlopers – whales, sharks, bony fishes and other sea life – go to reproduce, unpolluted, undisturbed coastal waters are needed on both sides of the tide-line. Yet the coastal zone has another important role to play – influencing climatic change.

During the process of photosynthesis, phytoplankton floating at the surface absorb carbon dioxide from the atmosphere, and this is trapped as carbon compounds in their microscopic bodies. When they die, the tiny cells drop to the sea floor and become part of the bottom sediment, making it an important carbon 'sink', or store. About 15 per cent of all the carbon dioxide produced by human activities is thought to be taken up in this way and buried in sediments covering the world's continental shelves.

This process establishes the continental shelf as an important contributor to the global carbon cycle, because carbon dioxide that is trapped in sediments is not in the atmosphere, and therefore is not contributing to the greenhouse effect (the way in which carbon dioxide traps the heat from the Sun close to the Earth's surface). This, in turn, has a major effect on patterns such as global warming, the melting of the polar ice caps and any consequent rise in sea levels.

The margins between land and sea therefore function as a sort of skin that is essential to the well-being not only of coastlines and their inhabitants but to that of the entire planet.

1

The Dynamic Coast

Damp Shore *Fog forms on the Oregon coast, bringing moisture to its temperate rain forests.*

Life never stands still in the zone where land meets sea. The very environment is dynamic as waves batter the shore, crumbling cliffs, grinding down the debris and throwing it back as beaches. Volcanic forces are at work – so are the wind-driven currents of the oceans. In some parts of the world, global warming poses a threat to the coral reefs that shelter many tropical shores. Hurricanes or typhoons, storm surges and tsunamis are among the more localised hazards of coastal life, wreaking periodic havoc. More benevolent are sea fogs, which roll inshore over the land and enable life to flourish even in the desert.

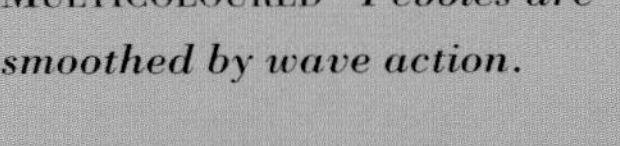

Multicoloured *Pebbles are smoothed by wave action.*

THE BATTLE BETWEEN LAND AND SEA

Flat waves push material up a beach; steep waves often push it down, eroding the beach. The relationship between land and sea in the zones where they meet is a delicate one, constantly affecting the shape of our coastlines.

A multitude of forces are at work shaping coastlines and continually reshaping them. From the majestic ocean inlets of fiords in Norway or New Zealand's South Island to the palm-fringed strands of tropical beaches, from the barely perceptible smudge of pancake-flat Dutch shores to the wave-battered crags of towering ocean-facing cliffs, the world's coasts display a dazzling variety, each stretch of them moulded in different ways through the interaction of different forces and circumstances. These forces range hugely in scale from the persistent action, over centuries, of tiny raindrops to the mighty clash of the vast plates that make up the Earth's crust. Rising sea level and sinking land are crucial factors; so, too, are agents of erosion including wind, ice, glaciers, rainwater and rivers. The nature of the rock or soil these forces act upon is another vital component. In England, along the coast of Yorkshire's Holderness peninsula, tipped by the spit of sand known as Spurn Head, the soft clays are eaten away by the sea at an average of more than 6 ft (2 m) a year.

Of all these influences, the movements of the Earth's crust are probably the most dramatic. As the theory of plate tectonics demonstrates, the Earth's crust is divided up into constantly moving plates, some colliding with one another, others very slowly pulling away from each other. Some coastlines are on the trailing edges of plates that are parting; others lie where they are colliding. Such regions, particularly the latter, are regularly shaken by earthquakes and volcanic activity.

The Red Sea, separating Africa from Arabia, marks a rift where plates are separating. It is part of a major gash, the Great African Rift, which runs from the Dead Sea in the north to Mozambique in the south, dotted along its length with recently active volcanoes, such as the cones marking the shoreline at Djibouti at the mouth of the Red Sea. The volcanic islands and deep ocean trenches of the Pacific are the result of a collision of plates where one plunges beneath its neighbour into the Earth's interior. The descending plate heats up and molten material rises to the surface as an arc of volcanoes behind the trench. Most of the world's deepest trenches are found along the margins of the Pacific, where some of the accompanying volcanoes break the sea's surface to form islands, such as the Aleutians and the Philippines, that are the dominant coastal feature. The islands of the Caribbean's eastern fringe were formed in the same way. They contain volcanoes like Mont Pelée on Martinique and Soufrière on St Vincent, which periodically spew searingly hot incandescent gases down their sides and into the sea.

Other islands have been formed over 'hot spots' in the Earth's crust. These are created by plumes of molten magma rising to the surface. As the crustal plate moves over the spot where volcanic activity is most intense, undersea mountains, called seamounts, are thrown up. When they reach the surface of the ocean they become volcanic islands. There are 16 recognised hot spots across the globe, including Hawaii, Easter Island and the Galápagos Islands in the Pacific; Iceland, Ascension Island and Tristan da Cunha in the Atlantic; and Réunion in the Indian Ocean.

SLIP-SLIDING AWAY *Coastal erosion threatens this North Carolina beach cottage.*

SUN, SEA AND ICE *Ice is the agent of erosion along the coast of the Antarctic Peninsula.*

> **FALLING INTO THE SEA**
>
> At Holderness, on the north-east coast of England, a 25 mile (40 km) stretch of cliffs has been retreating for 6000 years. In this period an estimated 185 sq miles (480 km^2) of land has been washed away, and since Roman times 36 villages are known to have been lost. In December 1996 a house in the village of Cowden had to be demolished to prevent it from falling into the sea.

The newest and most volcanically active island in the Hawaiian chain – Hawaii itself – is currently centred over the Hawaii hot spot in the mid-Pacific, but as the Pacific crustal plate moves westwards, at a rate of 3½ in (9 cm) per year, Hawaii will gradually over thousands of years become less active, and a new island, taking its place over the plume, will rise out of the sea to the east. Eventually Hawaii's volcanoes, such as Mauna Loa and Mauna Kea, will stop producing and depositing lava and ash, and erosion by the sea will start to take its toll. So, as the islands in the Hawaiian chain move away from the hot spot, they will slowly disappear below the waves, joining the Emperor Seamounts as great undersea mountain chains.

DROWNED VALLEYS

The agents of erosion that shape coastlines are not all marine in origin. In south-western England the 'crenulate' (finely notched) coast of southern Cornwall has a complicated outline of headlands, coves, islands and winding inlets and estuaries, called 'rias', all formed originally by rivers. The rivers carved valleys down to the sea, and these were later 'drowned' when the sea level rose after the ice ages. In south-east Australia the complex series of waterways that makes up Sydney Harbour similarly marks a system of drowned valleys.

HOT SPOT ISLANDS *A photograph from space shows the Hawaiian island chain, formed over a hot spot on the floor of the Pacific.*

SEA NOTCHED *The sea has invaded former river valleys in Sydney Harbour.*

Guyot

THE SEABED FROM SHORE TO DEEP SEA *Shorelines are found not only between the sea and major land masses but also around volcanic islands and atolls.*

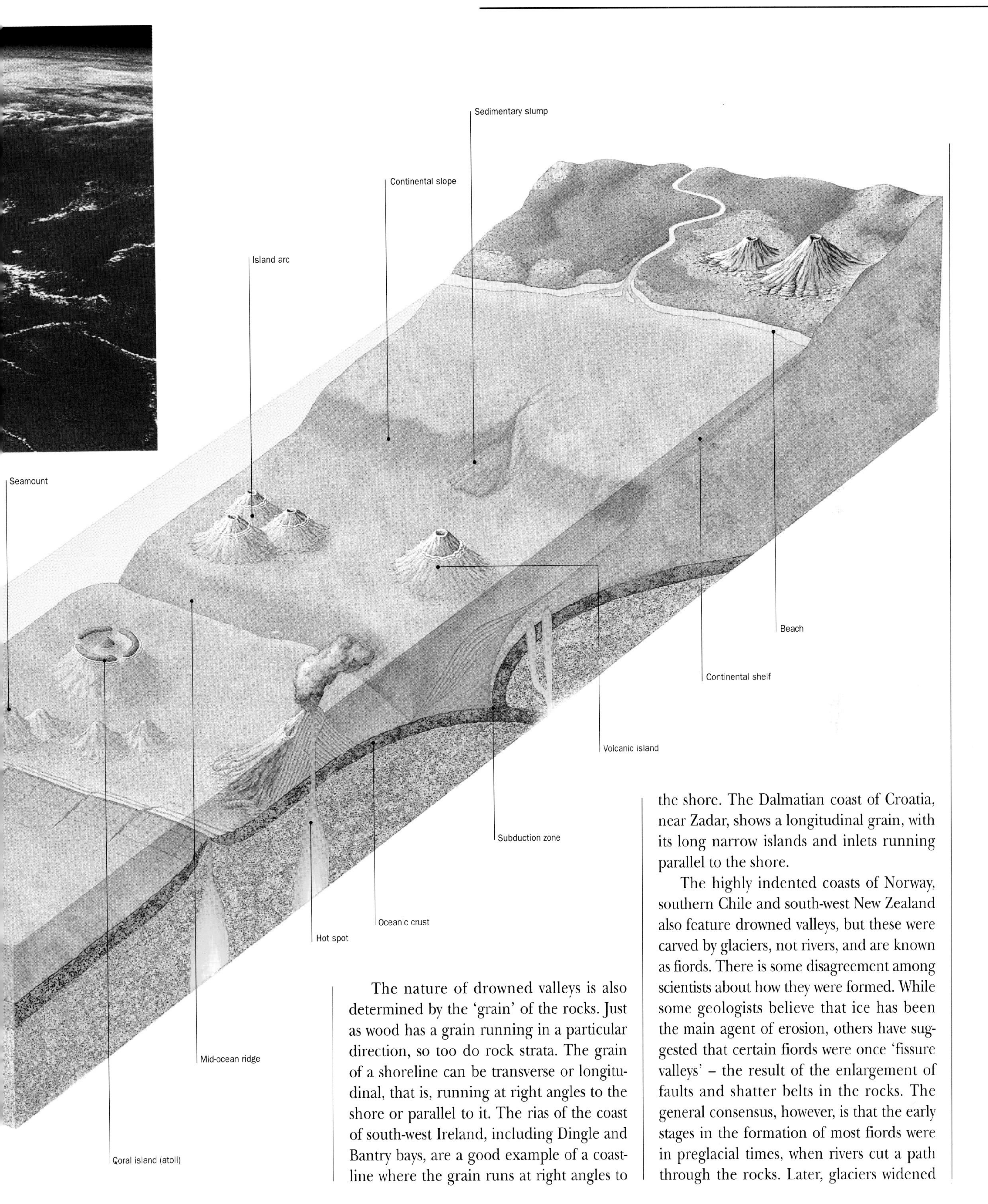

The nature of drowned valleys is also determined by the 'grain' of the rocks. Just as wood has a grain running in a particular direction, so too do rock strata. The grain of a shoreline can be transverse or longitudinal, that is, running at right angles to the shore or parallel to it. The rias of the coast of south-west Ireland, including Dingle and Bantry bays, are a good example of a coastline where the grain runs at right angles to the shore. The Dalmatian coast of Croatia, near Zadar, shows a longitudinal grain, with its long narrow islands and inlets running parallel to the shore.

The highly indented coasts of Norway, southern Chile and south-west New Zealand also feature drowned valleys, but these were carved by glaciers, not rivers, and are known as fiords. There is some disagreement among scientists about how they were formed. While some geologists believe that ice has been the main agent of erosion, others have suggested that certain fiords were once 'fissure valleys' – the result of the enlargement of faults and shatter belts in the rocks. The general consensus, however, is that the early stages in the formation of most fiords were in preglacial times, when rivers cut a path through the rocks. Later, glaciers widened

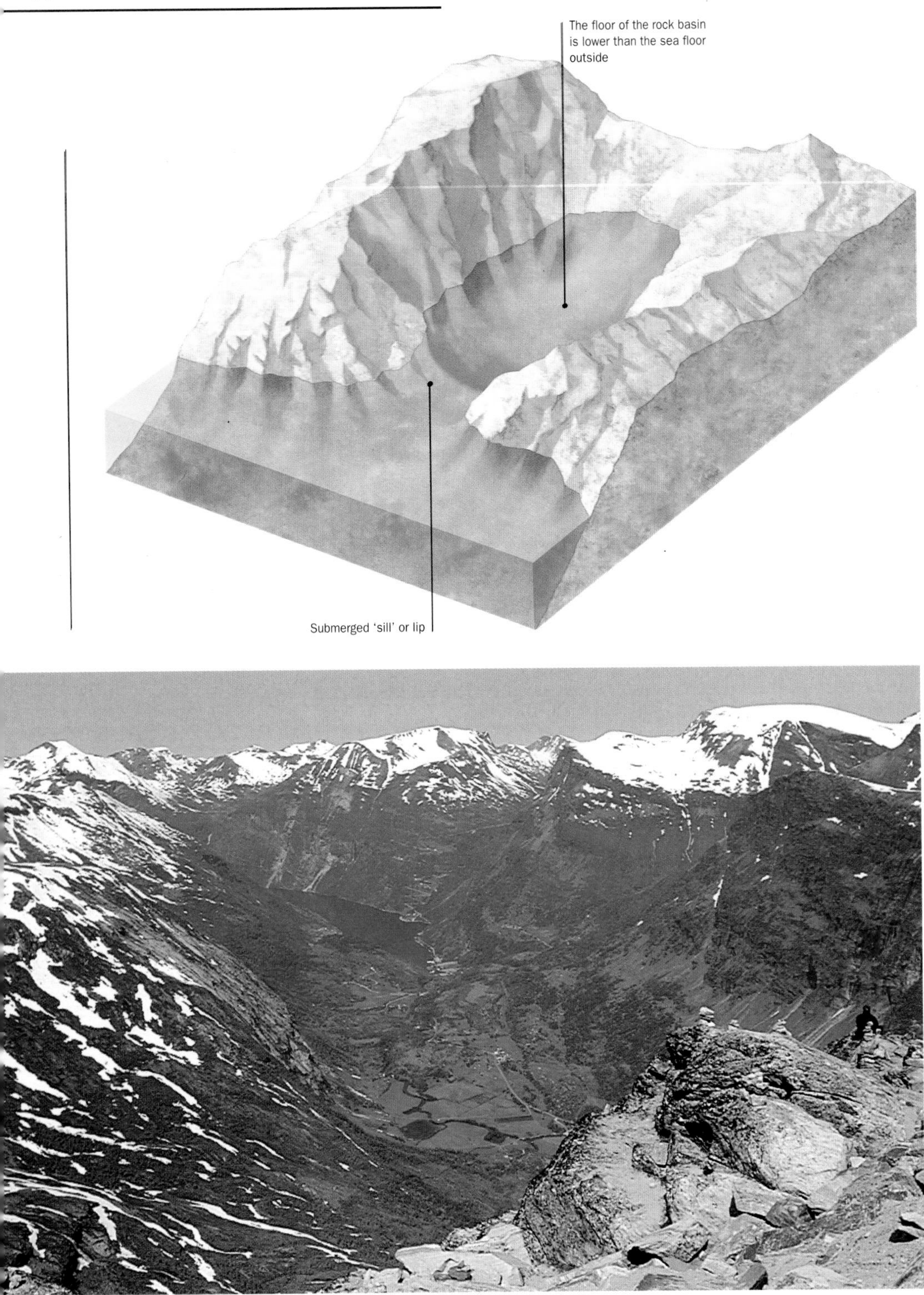

Ice Carving *A fiord (top left) has a deep basin and a shallow lip. Geiranger Fiord (bottom left) on the coast of northern Norway is a drowned valley carved by a glacier.*

and deepened the river valleys, changing them into true glacial troughs. In fact, the floor of a fiord is shaped like a trough and is usually lower than the sea floor outside.

Waves and Tides

The movements of the sea itself influence the appearance of a shoreline. The body of water moving to and fro during the ebb and flow of tides sets up tidal currents, which carry sediments into estuaries and inlets on the rising tide and move sand back and forth along a beach. Wherever their passage is restricted, such as round a headland or between islands, these currents can become violent 'tidal races'. The currents in a tidal race are strong enough to roll pebbles along the seabed and toss small boats about like matchsticks. Where the rise and fall of the tide is large, as in the Bay of Fundy on the east coast of Canada or the Severn Estuary on the west coast of England, the coast is said to be macrotidal. In the Bay of Fundy, a rise and fall of about 45 ft (14 m) represents the world's greatest tidal range. Coasts such as those of the Mediterranean and Baltic seas, by contrast, are subjected to only a small tidal range, and are described as microtidal.

Although tides and tidal currents help to erode or build a coast, by far the most important agents of destruction and deposition are waves. In tropical latitudes, where storms are less frequent than they are closer to the poles, waves are usually long, low and far apart, and tend to deposit materials. They are known as 'constructive' waves, building beaches – often long, sweeping ones with fine sand – rather than breaking them down. In temperate latitudes, on the other hand, waves erode or deposit according to the season. In summer there is a tendency for waves to build beaches as they do in the tropics, while in winter seasonal storms carry the beaches' materials away. These 'destructive' storm waves are common in higher latitudes, where they can destroy whole sections of coastline.

Whether deposition or erosion predominates depends on the nature of the 'swash'. Swash is the turbulent water that rushes up the beach after a wave has broken. Its upper limit is usually marked by a wet line in the sand, accompanied by accumulations of seaweed and driftwood. It can move material up or down the beach, depending on the nature of the wave. Flatter waves tend to push material up the beach, whereas steeper waves, which strike the beach almost from above, can push material either way. Some of the swash drains through the sand or shingle, the rest rushing back down the beach as 'backwash'. If backwash is stronger

Flower Pots *Marine erosion has created unusual rock formations in the Bay of Fundy, which has the greatest tidal range in the world.*

TIDES AND THE PULL OF THE MOON AND SUN

The pulse of the ocean is the rise and fall of the tide. The ancient Greeks thought that the tides were controlled by the moon goddess Diana, and in a way they were right. Tides are caused by the gravitational pull of the Moon and, to a lesser extent, the Sun. Put simply, as the Moon passes over the sea, its gravitational pull causes the sea to bulge, and the bulge passes around the planet as the Moon moves around the Earth.

The largest tidal ranges occur as 'spring tides' shortly after the time of new and full Moon. This is when the Sun and the Moon are both lined up with the Earth, exerting a joint gravitational pull that induces the highest tides. Relatively small tidal ranges, known as 'neap tides', occur near the time when the Sun and Moon are at right angles to one another, so that their different pulls to some extent cancel each other out. The highest and lowest spring tides of all occur at the time of the equinoxes, towards the end of March and September, when the combined pull of the Sun and Moon is at its greatest.

In reality, tidal movements are complicated by a number of factors including the various landmasses and deep oceans that block their flow. In enclosed bodies of water, such as the Baltic, the Mediterranean and the Black Sea, the tides are minimal. By contrast, in the Bay of Fundy on the east coast of Canada and the Severn Estuary on the west coast of Britain, water levels can rise and fall by as much as 30-45 ft (10-14 m) twice each day. If there is just one high water and one low water during the tidal day, it is known as a diurnal tide. If there are two low waters and two high waters during the tidal day, as experienced, say, in the English Channel, it is called a semidiurnal tide.

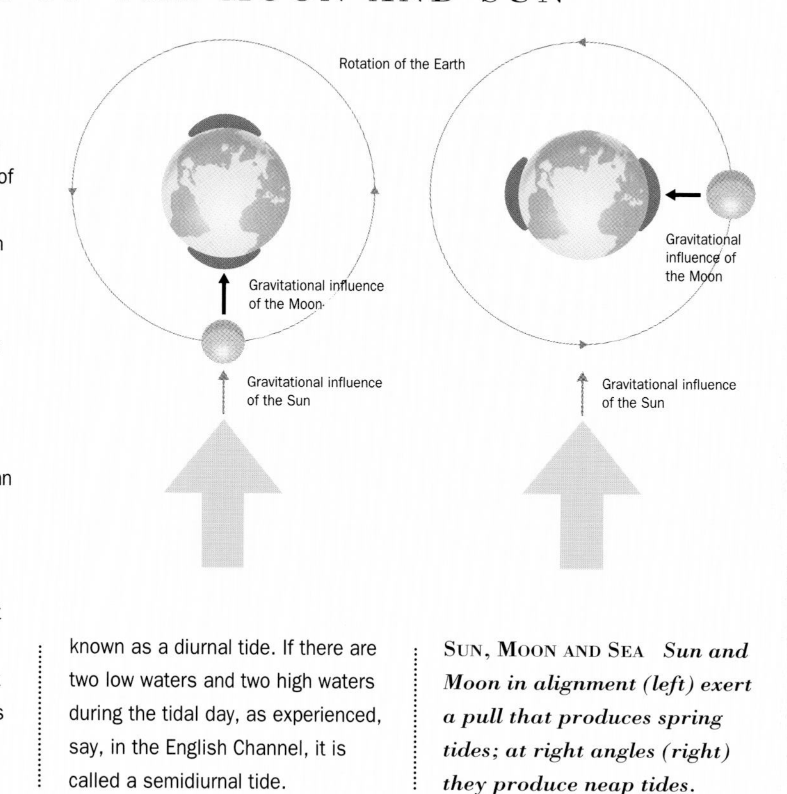

SUN, MOON AND SEA *Sun and Moon in alignment (left) exert a pull that produces spring tides; at right angles (right) they produce neap tides.*

than swash, as it is in destructive waves, the net effect is to remove material from the beach.

Where waves break against cliffs during a storm, they exert tremendous pressures. Stones and boulders carried by the sea and hurled at the cliff-side help to break down the rock, a process known as attrition. Thus waves – together with rock-boring marine organisms, rain, wind and chemical weathering by the seawater itself – will attack the lower part of a cliff, slowly undermining it until the upper part tumbles into the sea. The new cliff, standing farther back than the old one, will have a ledge known as a 'wave-cut platform' at its base. If the sea level has been relatively stable the platform can be quite extensive, with a gradual incline of about 1 in 100 into the sea. Examples are found all over the world, including the Oregon coastline in North America.

Where substantial areas of hard rocks alternate with large areas of softer rocks, a coastline of headlands and bays may be formed, such as that of Brittany in northwestern France. The headlands, often of tougher rocks such as granite, take the brunt of the wave action, and the materials eroded from them are deposited in the intervening bays. When the processes of erosion and deposition have taken things to their logical conclusion, the result is a straightening of the shoreline, as along a massive 750 mile (1200 km) stretch of the Great Australian

PARADISE BEACH *Willies Creek in Western Australia has sun, sea, and sand deposited by the Indian Ocean.*

TWELVE APOSTLES *Wave action has eroded the rocks along the Port Campbell coastline of Victoria, Australia, leaving spectacular sea stacks (top). This natural arch (above) was carved out by the sea on Italy's Puglia coast.*

Bight where the huge limestone plateau of the Nullarbor Plain ends at the sea in sheer white cliffs, some stretching unbroken for more than 125 miles (200 km).

HIGH COASTS

The dazzling white chalk cliffs of Beachy Head and the Seven Sisters on the south coast of England rise 534 ft (163 m) above the waves crashing at their feet. The yellow-and-brown cliffs at Moher in western Ireland reach 670 ft (204 m). West of Melbourne in Australia is another spectacular coastline, stretching for 20 miles (32 km) and now protected as the Port Campbell National Park. Here cliffs and jagged limestone islets rise from the sea in all kinds of shapes and sizes. All these shorelines were sculpted by the abrasive activity of waves and wind.

Vertical cliffs, composed of hard, resistant rocks, usually occur on exposed coasts where wave action is vigorous. They may be accompanied by offshore stacks, arches, wedges and chimneys – the remnants of cliffs that disappeared long ago. In northern France the Porte d'Aval arch and L'Anguille rock needle are the result of the sea cutting into resistant chalk cliffs. Similar rock formations are found along the Oregon coast of North America and on the south coast of

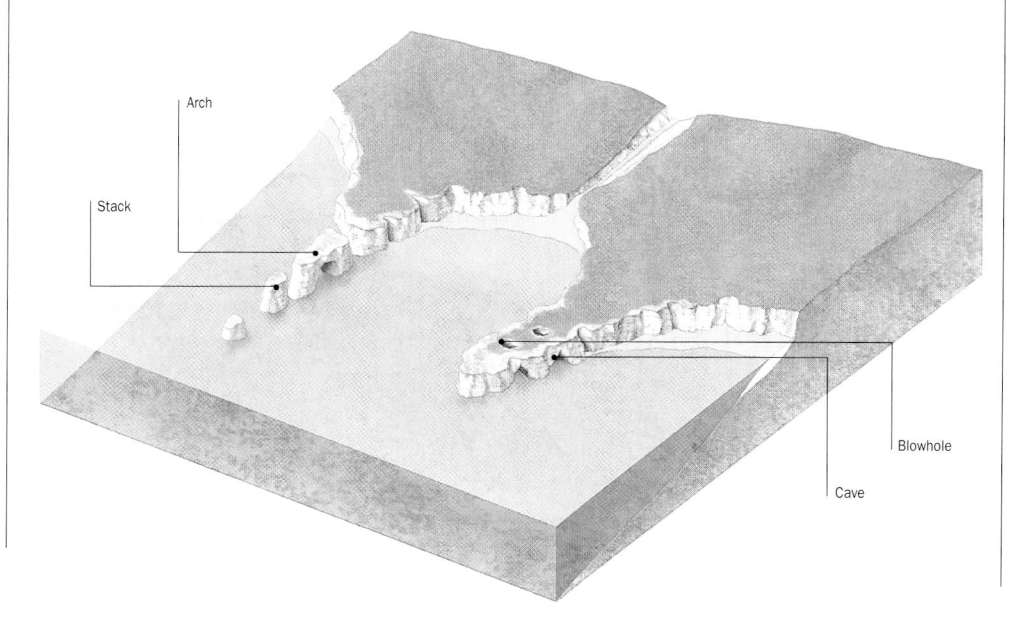

HIGH COAST *The sea has eroded soft rock to form coves, leaving hard rock as headlands. Further erosion forms arches and stacks.*

Explosion of Water *The sea gushes through a blowhole (left) on a rocky Galápagos island shore. Six-sided basalt columns rise on either side of Fingal's Cave (above) on the Scottish island of Staffa.*

England at Durdle Door, near Lulworth Cove in Dorset, where a white limestone arch forms the tip of a headland.

At Port Campbell the rock stacks known as the Twelve Apostles stand as silent witnesses to an ancient shoreline. The limestone rocks, laid down 26 million years ago, have been eroded, great slabs crashing down from the cliffs into the sea every 20-30 years. Where patches of soft rocks were eroded sooner than the harder rocks above and on either side of them, arches have formed. One such Port Campbell formation, known as London Bridge, was until recently a flat-topped headland with two arches linked directly to the cliffs, but is now an island with one arch. The other crashed into the sea in 1990, stranding a couple of tourists who had to be rescued by helicopter.

Along the same stretch of coast, large 'blowholes' have formed where the sea has worn away softer rock strata and bored long tunnels into the cliff-face by attrition, chemical weathering and wave action. The roof at the end of a tunnel eventually collapses, allowing the water inside to be pushed up to the cliff top. At Port Campbell, the Blowhole is almost 1/4 mile (0.4 km) long. In other parts of the world, as on the rocky coast of Christmas Island in the Indian Ocean, whole rows of blowholes along the shore spout like decorative fountains each time waves race up the tunnels. The water is compressed by the narrowing boreholes, and is thrust high into the air in pressurised jets.

Similarly, grottoes and sea caves are formed when the sea attacks cracks and other weaknesses in the rock. The most famous sea cave is probably Fingal's Cave, on the small uninhabited island of Staffa off the west coast of Scotland, where the fissures between six-sided columns of tough basalt rock have been opened by the waves. The result today is a series of spectacular caves, bounded by tall, cathedral-like arches of basalt. Fingal's Cave, which inspired the German composer Mendelssohn to write his *Hebrides* overture, is the largest of these. It extends 230 ft (70 m) into the rock and has a cave mouth 65 ft (20 m) high.

Cliffs composed of softer rocks are called 'drift' cliffs. These are less steep and are particularly liable to slumping and landslides. On the Dorset coast of southern England, cliffs of clays and marls are a gift to fossil hunters. After a heavy south-westerly storm the waves often cause chunks of the cliffs to collapse. The backwash pulls out debris, including the fossilised bones of dinosaurs and ichthyosaurs. These remains can be picked up on the beach.

BUILDING BEACHES

Waves create beaches – and destroy them. In the never-ending to and fro as waves break along the shoreline and then withdraw again, sediments – pebbles, shingle, sand and even finer particles – are constantly being tugged out to sea, washed back onto the coast and then back out to sea again. Beaches form where the waves are relatively gentle, depositing more than they take away; higher-energy waves take away more than they deposit. The sediments will usually have been washed into the sea by rivers or along the coast by currents. On the whole, sandy beaches are more common in lower and middle latitudes, shingle beaches in higher latitudes (those closer to the poles). This again is partly because of the waves; in higher latitudes the wave action is often more turbulent and therefore capable of transporting heavier material, such as pebbles and shingle.

LAND LINK *A shingle bar, or tombolo, joins Pepin Island to the mainland at the north of New Zealand's South Island.*

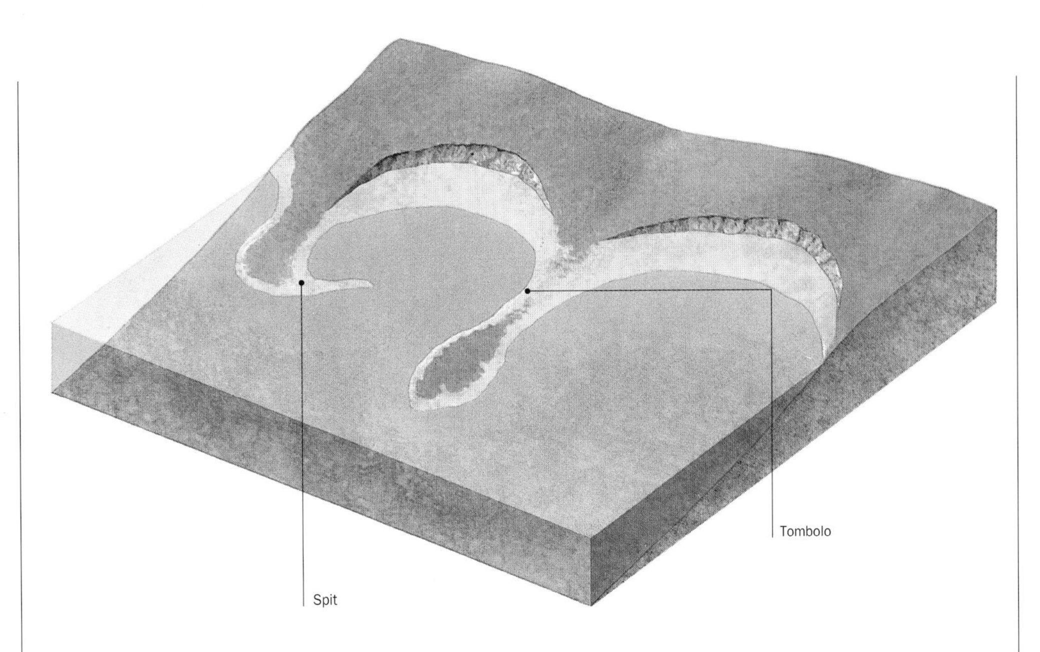

COAST DEPOSITS *Tombolos and spits form where sand is thrown up by the sea. On the Devon coast, south-west England, hard and soft rock alternate to form headlands and bays.*

The way in which the materials are deposited will depend on the direction of the wind, waves and water currents. Beaches tend to accumulate in sheltered bays, but a headland can have a beach too. If the line of the coastline suddenly changes direction, narrow ridges of sand or shingle, known as spits, can form. On the Suffolk coast in eastern England a great shingle spit has formed at Aldeburgh and diverted the River Alde south for about 10 miles (16 km) before it opens to the sea. The growth of this spit is well documented in historical records. It was 5 miles (8 km) long and had already reached the village of Orford when a castle, its ruins still standing, was built there by Henry II in 1165. By Elizabethan times the spit had advanced to the south by 2 miles (3.2 km), and by the 18th century it had built up a further 1/4 mile (0.4 km). When Ordnance Survey maps first appeared in 1805, the spit had reached its farthest point opposite the seaside village of Shingle Street. During the next 50 years or so it receded 2 miles, but then it advanced south again so that its tip, known as North Weir Point, now lies about 3/4 mile (1.2 km) from Shingle Street. On the river side of the spit there are extensive salt-marshes, while the seaward side is a long beach which continues north past Aldeburgh, almost uninterrupted, all the way to the bay of The Wash.

The processes of deposition can also fill in the gap between an island and the nearby mainland to form a beach known as a 'tombolo'. Good examples are found at Marble Head Island, Massachusetts, and Chesil Beach, England, linking the Isle of Portland to the Dorset mainland. Deposition can also lay down material across river and harbour mouths and bays to form bars.

The movement of beach materials along a coast is known as longshore drift. Waves

Glistening Pebbles *Beaches in higher latitudes tend to be composed of pebbles and shingle like this one near New Aberdour, Scotland.*

are the main agents propelling material along the shore, the swash pushing sand and shingle obliquely across the shore, while the backwash pulls it straight down to the sea. In this way, beach materials progress across a beach in a zigzag fashion. The distances involved can be considerable. On the south coast of Britain, for example, recognisable quartzite pebbles from Budleigh Salterton in south Devon have been discovered on the beaches of Kent and Sussex, more than 200 miles (320 km) farther east.

Any interference with this natural movement can have quite devastating effects. At Folkestone in Kent, the harbour regularly filled with shingle until the western pier was extended and the shingle piled up on its western side instead. The problem of shingle in the harbour was solved, but to the east of the town beaches were unprotected, allowing the waves to pound directly against cliffs in the area of the Warren, causing large tracts to slump and fall into the sea.

Sandy Shore *Long, low waves deposit sand on a beach on Morocco's Atlantic coast north of Essaouira.*

Low Coasts

Low coasts, usually of softer rocks, are more easily moulded by the sea than high coasts. The result is often a fairly straight coastline, where cycles of erosion and deposition can operate over long stretches of shore. Such coasts are usually sandy by nature and consist of barriers of sand, sand dunes, lagoons and salt marshes. The eastern and Gulf of Mexico coasts of the USA are typical of this kind of coastline, where sandy barrier islands, topped with sand dunes and backed by shallow lagoons, separate the low-lying mainland from the extensive open-ocean beaches on the seaward side of the barrier.

Sand dunes form behind these low coasts where the wind-blown sands are trapped by plants to form small mounds. They build up in three phases: first the young foredune, then the main mobile dune and finally the mature stabilised dune. The foredune is closest to the sea and grows to about 10 ft (3 m) high. It is characterised in temperate latitudes by sand couch grass (*Agropyron junceum*), which tolerates being doused in seawater. Mobile dunes have a cover of marram grass on the slopes that face inland, but the slopes that face out to sea are still bare, so the sand is susceptible to being blown by the wind. These dunes can be immense; those at Culbin Sands on the east

OPEN SANDS AND BEACHES THAT WHISTLE AND BARK

Quartz is the hardest and most common mineral found on mainland sandy beaches, and it accounts for the colour and texture of the classic 'golden sands'. On oceanic islands quartz is rarer, and beaches of greenish or even black volcanic sand are more common. In the Hawaiian islands it is mainly basaltic, alkaline volcanic rocks that contribute to beach deposits. Some Florida beaches have spherical glass shards which have floated, with the aid of gas-filled central cavities, from volcanoes in the West Indies.

Pink or white sand beaches contain fine fragments of coral or shells. The pieces can be large and sharp, and may cut the feet of people walking over them. If the shells have been ground down, however, they cling to the skin and are difficult to remove. They come from the remains of shellfish, such as cockles and razorfish, which are buried below the sand close to the low-tide mark. Their numbers can reach thousands per cubic foot, and so over thousands of years the shells can be thrown up on the shore to build entire beaches out of their debris, such as Shell Beach on Alderney in the Channel Islands.

Some sandy beaches make distinctive noises. At Porth Oer ('Whistling Sands') on the Lleyn Peninsula of North Wales, the sands emit a soft squeaking sound when walked over during dry weather. The noise is created when the very fine sand particles are rubbed together. The 'barking sands' of Kauai, one of the Hawaiian island chain, are on a black-sand beach that also squeaks when you walk over it.

ARRAY OF SANDS *Black basaltic sands on Hawaii (above), sea-smoothed coral fragments in the Caribbean (far left) and shell fragments on Florida's Gulf coast (left) – beaches are as varied as their constituents.*

coast of Scotland – the largest in the British Isles – reach a height of 98 ft (30 m). The mature stabilised dune is protected from the wind by the mobile dunes in front and is fully covered with vegetation, preventing the sand from being blown away.

Sand dunes can be very ancient. The Skeleton Coast of the Namib Desert, on the Atlantic shore of south-west Africa, may have the oldest in the world – about 130 million years old according to some geologists. Most researchers agree that conditions have changed very little for the past 80 million years, and a fossil desert of ancient dunes about 40 million years old has been found

LOW COAST *While a salt marsh forms on the leeward side of a sand spit, the windward coast opposite is backed by sand dunes. Left: A bank of cockleshells has been thrown up by the sea at the edge of a salt marsh.*

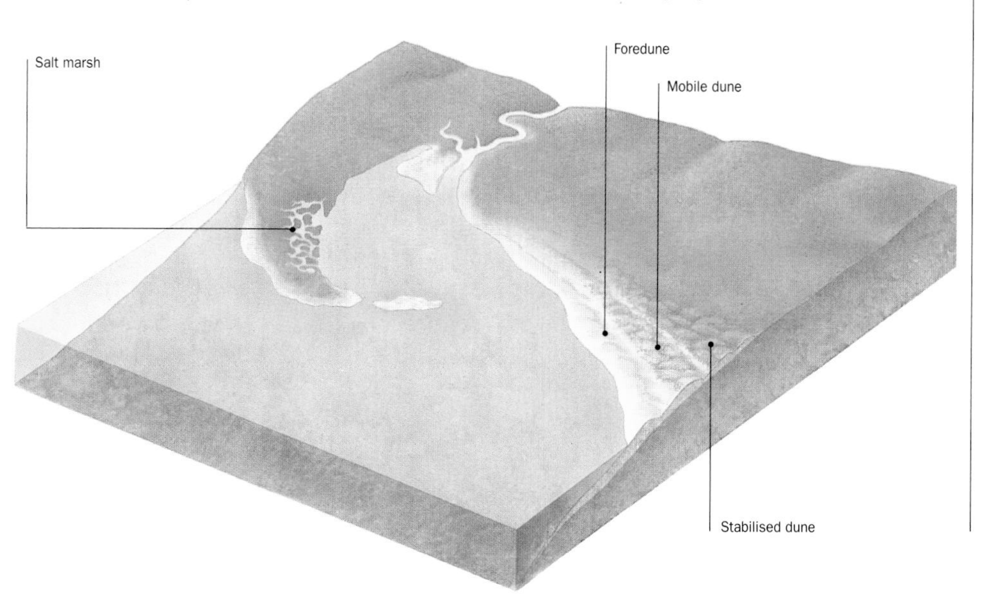

In the Dunes *Namibia's Skeleton Coast (above) is host to an entire community of desert animals. Ragwort and marram grass are becoming established in the shifting sand of a European dune (left).*

below the present dune fields between Kuiseb and the Orange River. These ancient dunes, like their present-day counterparts, were made of sediments carried down from the Lesotho Highlands by the Orange River. The sands were washed out to sea and then back to the shore.

On so-called 'organic coasts', where wave energy is at a minimum – such as in deep, protected bays – sediments accumulate. This gives rise to salt marshes in high and middle latitudes, while mangroves grow in tropical regions. The dense vegetation is able to trap silt and sand brought in by the tide. Reef-building corals also form organic coasts, building up fringing reefs that protect the coast of the mainland.

Coastal Plains

Where the land at the margins of continents slopes down to the sea at a gradient of less than 1 in 100, it is known as a coastal plain. Coastal plains can be either erosional or depositional, and the forces of rivers, glaciers and the sea itself have a role to play in creating both.

In Scandinavia the 'strandflat' of western Norway is an erosional plain; it was created by glaciers descending from the Norwegian highlands to the Atlantic coast where, many years ago, they scraped the rocks almost flat to form a broad, low-lying, gently sloping coast with ice-smoothed outlying islands.

Depositional coastal plains include river deltas, the largest of which are created by rivers such as the Mississippi which carry sediments – mostly fine muds and silts – in suspension. As the river water is lighter than the salt water of the ocean, the sediments are carried many miles out to sea. As the flow of the river slows down, and its influence widens, the sediments are deposited on the sea floor. The coarser material drops out first, so the sediments become finer farther away from the river mouth.

About 65 million years ago the Mississippi delta lay some 500 miles (800 km) farther north, around what is now Cairo, Illinois. About 8000 years ago the river delta was located around modern Baton Rouge, Louisiana, now 62 miles (100 km) inland. Since then the river has created 11 031 sq miles (28 568 km²) of delta, of which 9299 sq miles (23 900 km²) are above water – though these are now under attack because of the rising sea level. The delta does not just advance and retreat; it also moves laterally. The Mississippi river alters its course every 1000 years or so, and American geologists have identified several abandoned deltas. The youngest of them is at Lafourche, which was abandoned by the river about 300 years ago.

Not all deltas are formed simply by the processes of deposition; the deltas of the Rhône and Ebro in the Mediterranean have also been shaped by the action of waves and tides. The sea has swept the river-borne sediments to the side, creating beaches and spits, and the sheltered areas behind have become marshes and swamps, as in the Camargue in southern France.

Some large rivers have no deltas at all. At the mouth of the mighty Congo or Zaire river, for instance, a huge submarine canyon extends 15 miles (25 km) into the ocean. Sediments are flushed from the river mouth by repeated floods and underwater avalanches, known as turbidity currents. The flow of these can be fast and powerful, reaching speeds of 50-60 mph (80-100 km/h). The current digs out the sides and floor of the canyon and carries the fine particles out to the deep ocean floor beyond. The result is a submarine fan of sediment on the seabed at the mouth of the canyon.

The Continental Shelf

Fringing the continents below the waves is the continental shelf, a gently sloping underwater extension of the continent. It can be wide or narrow. The shelf of western Europe contains the entire British Isles, as well as underlying the North Sea, Skagerrak, Baltic Sea, Irish Sea and English Channel.

The features of a continental shelf are generally similar to those of the adjoining coastal plain, although tides and currents modify the surface somewhat. Sediments accumulate on it, their weight causing the shelf to sink so that it remains below sea level. The strata formed on a continental shelf tend to trap beds of organic materials, and so are the focus for oil exploration. The shelf usually has an abrupt edge where the sea floor drops away steeply into the abyss. This marks the seaward boundary of the place where land and sea meet.

Ol' Man River *An infrared photograph taken from space shows the bird-foot shape of the Mississippi delta and the flow of sediment out to sea.*

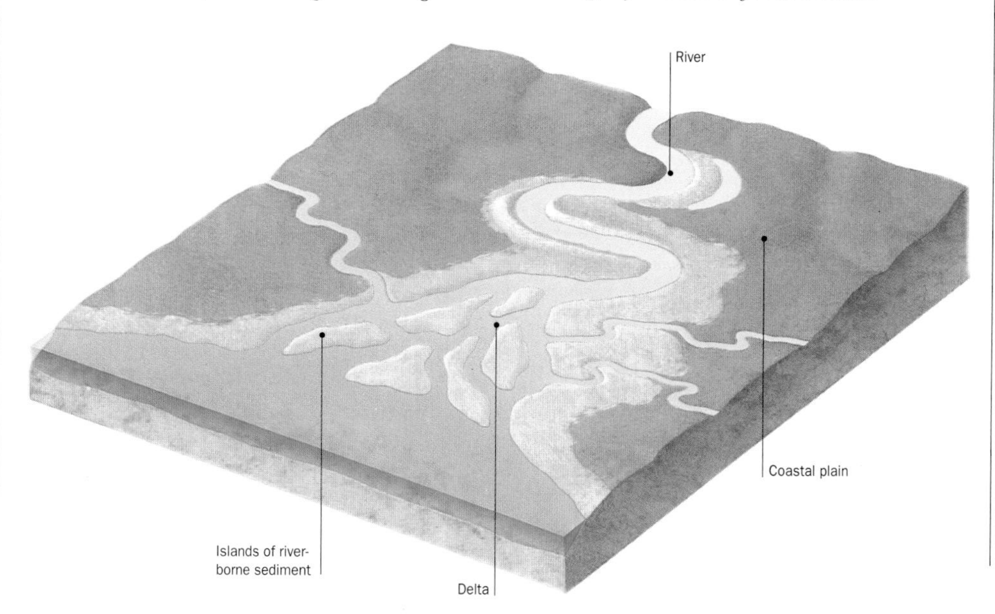

Coastal Plain *A river criss-crosses the coastal plain and deposits silt in the river mouth at the edge of the sea to form a delta.*

Atmosphere and Ocean

The coastline equation includes a third element, as well as land and sea: the atmosphere. Global warming, wind-driven ocean currents – events in the atmosphere affect everything from the growth of coral reefs to the wealth of coastal feeding grounds.

Prevailing winds drive the surface currents of the world's oceans, and the complex interplay of the different currents has a crucial impact on life in the zones where land meets sea. Thanks to the cold Humboldt Current, penguins normally associated with the Antarctic and the chilly archipelagos of the Southern Ocean thrive in the Galápagos Islands, which actually straddle the Equator. Cold water welling up from the ocean depths to the surface – off the coast of Oregon between April and October, for example – brings with it nutrients that allow the tiny drifting plant-like organisms known as phytoplankton to 'bloom'. They are consumed by animal plankton (zooplankton), which are eaten by small fishes; these, in turn, are eaten by larger fishes, and so on up the food chain. Where plankton abound, so do all other forms of marine life. Fish and squid are so abundant off these coasts that huge numbers of sea birds and sea mammals converge there to feed and breed.

Rich ecosystems like these are easily affected by changes in the conditions that shape them, however. Reports from all over the world have shown the effects of global warming on coastline plant and animal communities. In recent years, triggerfish (*Balistes carolinensis*), common stingrays (*Dasyatis basinica*) and electric rays (*Torpedo nobiliana*), more usually restricted to warmer waters to the south of the Bay of Biscay, have been found trapped in lobster pots off south-west England or brought to the surface in fishing trawls. A small but significant rise in the average temperature of coastal waters off Jamaica in the Caribbean is causing the local coral reefs to 'bleach' – which means, in effect, that they are ceasing to grow, thus threatening the diverse communities of plants and animals that depend upon them.

The Winds that Blow

Each latitude has its prevailing winds. The tropics have the trade winds; the mid-latitudes in both hemispheres have their westerlies; the polar latitudes the easterlies. Driving currents such as the Oyashio, which bathes Japan, and the Agulhas Current, which brushes the south-east coast of Africa, these winds play a leading role in the drama that unfolds where air, sea and land interact.

Winds control the currents in a very precise and predictable way, a phenomenon known as Ekman flow, after the Swedish oceanographer Vagn Walfrid Ekman (1874-1954), who studied the dynamics of ocean currents. The wind pushes the surface of the sea

GUANO GROUND *Humboldt penguins in south Peru stand on compacted layers of sea-bird droppings accumulated over hundreds of years.*

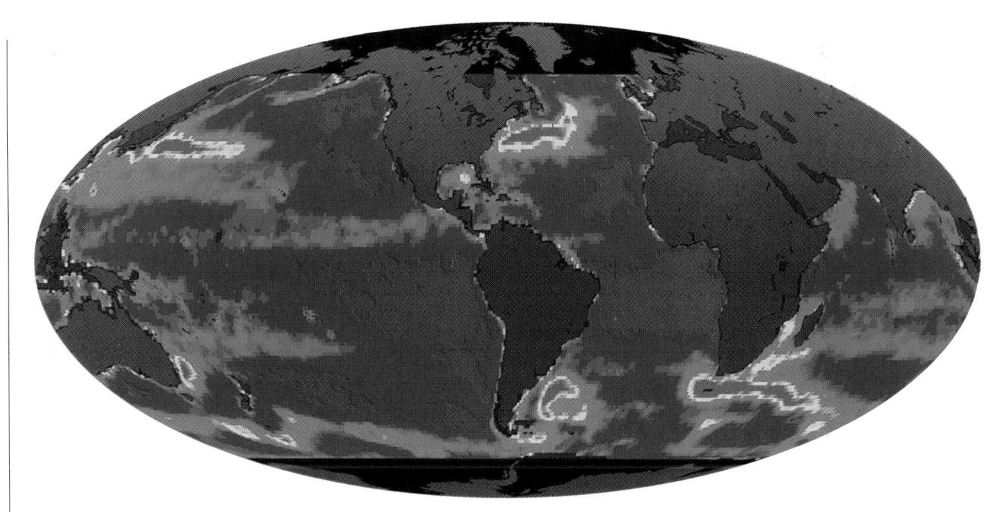

Ocean Currents *Currents in the Atlantic and Pacific oceans flow clockwise in the north and counter-clockwise in the south. The Gulf Stream off Florida and the Kuroshio off Japan show up red on the satellite photograph (right).*

by friction in a particular direction, but only the top 330 ft (100 m) or so – the Ekman layer – is moved. And even at the surface, the ocean's response is not simple, for other forces are operating, too. The currents driven by the wind are deflected by the so-called Coriolis effect – named after its discoverer, the French mathematician Gaspard Coriolis (1792-1843). This is a result of the rotation of the Earth and means that in the Northern Hemisphere currents are deflected by about 45° to the right of the wind, while in the Southern Hemisphere they are deflected by about 45° to the left.

Thus the Gulf Stream in the North Atlantic follows a curved path, moving north along the east coast of the United States, north-eastwards across the Atlantic, down the western coast of Europe to the Equator, from where it returns, via the east-flowing North Equatorial Current, to its starting point in the Caribbean. The Gulf Stream carries reserves of heat built up in the sun-soaked Caribbean and transports them across the Atlantic, bringing warm water and mild winters to north-west Europe. Its northerly branch – the North Atlantic Drift – carries the warmth even farther to the north so that Iceland, Spitzbergen and Jan Mayen Island are warmer than their Arctic or near-Arctic positions would suggest. The Gulf Stream is an impressive water conveyor, transporting up to 30 billion gallons (135 billion litres) every second – a staggering 65 times as much as is carried per second by all the rivers in the world.

In the South Atlantic the trend is for currents to be deflected to the west. The northward-flowing Benguela Current follows the coast of West Africa and veers to the west, crosses the Atlantic as the South Equatorial Current, bends to the south as the Brazilian Current, and then travels back across the South Atlantic.

Ekman flow is thus an important phenomenon which has several far-reaching effects on certain coastal waters. Along the Pacific coasts of North and South America, and the Atlantic coast of Portugal, north-west Africa and south-west Africa, for example, the wind blows roughly parallel with the coast, pushing aside the surface layer in the main ocean currents that flow alongside these shores – the California Current in the north Pacific, the Humboldt Current in the south Pacific and the Benguela Current in the south-east Atlantic. Deeper, colder waters, known as 'upwellings', rise up to replace the displaced surface waters. They come up from depths of as much as 1000 ft (300 m) and form a band 12 miles (20 km) wide along the shore. The temperature of

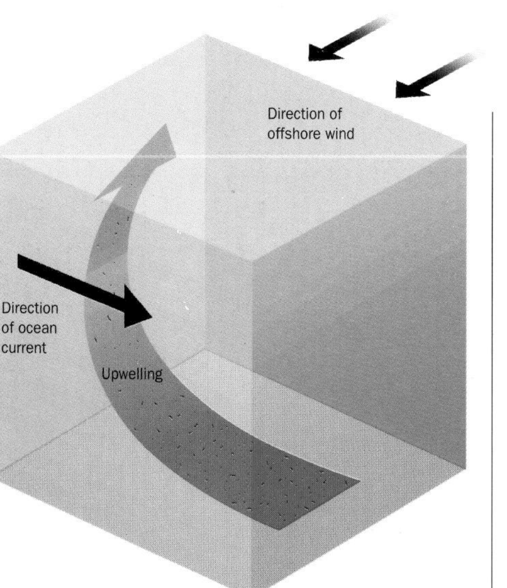

Upwellings *Offshore winds blowing across ocean currents close to land cause nutrients from the seabed to rise to the surface. A photograph from space (right) shows a large spiral of phytoplankton bloom (pale blue) off the New Zealand coast.*

the sea off Peru and northern Chile, where the upwelling is greater than anywhere else in the world, is 5-10°C (9-18°F) lower than the average for those latitudes.

Cold Productive Waters

Upwellings encourage rich growths of phytoplankton, as is proved at the Equator, another site of upwelling. In 1978 a Nimbus-7 satellite with a coastal zone colour scanner on board was put into orbit; it beamed back pictures of a patchwork sea of blues and greens according to the amount of phytoplankton present. The photographs showed the Equator as a coloured stripe, as if somebody had placed a marker pen on the spinning globe, indicating high levels of phytoplankton activity there.

This richness is the result of easterly winds blowing along the Equator which cause the surface waters to diverge. Waters to the north of the Equator move north, while those to the south flow in a southerly direction. Deep water, rich in nutrients, rises to take their place, and the phytoplankton blooms.

Elsewhere in the world, the Southern Ocean, which is the only ocean that flows

uninterruptedly around the entire globe, shows concentric rings around the Antarctic containing concentrations of phytoplankton. In high northern latitudes winter storms and currents, rather than upwellings, stir up the bottom sediments, followed each spring by an extraordinarily widespread bloom of phytoplankton, the largest biological event on the planet.

Phytoplankton are vital organisms whose importance spreads well beyond the particular seas where they are found. As well as forming the first link in the ocean food chain, they act as a 'sink' (store) for carbon dioxide – they absorb it and thus take it out of the atmosphere.

Phytoplankton may also be responsible for increasing the cloud cover over the Earth. Researchers from Australia's Commonwealth Scientific and Industrial Research Organisation (CSIRO) have sampled the air above inshore waters at Cape Grim off Tasmania and discovered a correlation between cloud condensation nuclei (the tiny particles on which water droplets condense to form clouds and fogs) and a chemical called methanesulphonate. This is an oxidised form of another compound with a tongue-twisting name, dimethylsulphide, which is a waste product of phytoplankton. When the sun blazes down on the sea, the phytoplankton are stimulated into reproducing. As the numbers grow, the tiny plants produce their waste chemicals and these give rise to more clouds. The clouds reduce the amount of sunlight reaching the sea, and the phytoplankton slow down their rate of growth. With less algae, less waste products, and therefore less cloud nuclei, cloud cover is reduced and more sunlight reaches the surface of the sea once more. Some scientists believe that the phytoplankton may act as a regulating system, countering global warming to some extent.

Fog in the Desert

Another effect of upwellings is to bring frequent coastal fogs, which allow a surprising variety of life to thrive in places such as the deserts that line the coasts of southern Peru and northern Chile. At night a light, warm breeze from the land blows towards the cold sea, and the fogs form. By day the sea breeze blows the fogs back onto the coast. In Peru a thick mist known as the *garua* masks the sun for long periods between June and December. As far

VARIABLE FLOW

Ocean currents at different depths flow in different directions and at different speeds. The warm surface waters of the northward-flowing Gulf Stream pass Cape Hatteras, North Carolina, at 40 miles (65 km) per day. Near the seabed, by contrast, a mixture of northward and southward-flowing currents of dense, cold, salt-rich 'bottom waters' – originating under the Arctic and Antarctic sea ice – move 10 miles (16 km) per day at most. At 4900 ft (1500 m) deep, warm, saline 'mid-waters' push westwards.

Plankton Pastures *These minute organisms, viewed here under a microscope, form the basis of the ocean food chain.*

Where Desert Meets Sea *Sea fogs bring moisture to the Atacama Desert of Peru and Chile.*

as the slopes of the Andes, more than 60 miles (100 km) inland, plants rely for survival on the water in the mists and fogs condensing on their stems and leaves. The water droplets then drip from the vegetation to the ground and are absorbed through the plants' roots. One coastal sand-dune plant, *Tillandsia*, absorbs the water through its leaves.

Similar conditions occur in south-west Africa, where the northward-flowing Benguela Current bathes the Skeleton Coast at the edge of the Namib Desert. The fogs of the Namib, which occur on about 60 days per year, are formed at night when moist air from the South Atlantic passes over the cold waters of the Benguela Current. The moisture in the air condenses, the droplets adhering to dust particles to form mist or fog. This rolls inland for about 60 miles (100 km), providing the main source of water for the plants and animals of the desert.

Refreshing Fog *A sea fog forms at the edge of the Namib Desert (below). A Namib beetle (right) drinks the condensation that forms on its body from sea fog.*

A Change in Temperature

Global warming is having a dramatic effect on plant and animal communities throughout the world, and the most obvious signs are in the coastal waters, in both temperate regions and the tropics. On the margins of the Pacific Ocean, Monterey Bay, California, is one of the world's best-studied coastlines. Here sea temperatures are reported to be 0.75°C (1.35°F) warmer than they were 60 years ago – a minute rise but with significant consequences. The plants and animals

POOR VISIBILITY: FOGS AND MISTS

Mists and fogs are common features in coastal regions. Where land and sea meet, temperature changes can be abrupt, providing just the right conditions for water vapour to condense as water droplets in the atmosphere. When visibility is reduced to less than 3280 ft (1000 m), the result is known officially as a 'fog'; visibility greater than this is called a 'mist'. Dust or pollution in the atmosphere creates a 'haze'. Fogs fall into three general categories: advection or sea fog, radiation fog and sea smoke. Advection fogs form when warm, moist air encounters a colder sea surface. They can be extremely dense, sometimes hundreds of feet thick. They occur in summer over sea-ice in both the Arctic and the Antarctic. The fogs on the coasts of California, Peru and Namibia are also advection fogs, caused by warm sea breezes crossing the cold waters associated with upwellings. The prevailing westerlies of the mid-latitudes mean that advection fogs generally form on the western coasts of continents in the same latitudes.

Radiation fogs appear when warm moist air comes into contact with ground that has cooled, for example overnight. On the coast they are more common in estuaries. They appear at night and are soon dispersed by the warmth of the rising sun.

Sea smoke forms when cold air passes over a sea surface that is 10°C (18°F) warmer than the air. Steam appears to rise from the surface of the sea, a phenomenon often seen in the Arctic and Antarctic in autumn and winter.

Fogs generally are more common in high and middle latitudes. In the North Atlantic dense fogs are frequently encountered off the coasts of Labrador and Newfoundland, where the cold Labrador Current encounters the warm Gulf Stream. In the North Pacific most fogs occur to the south of the Kamchatka Peninsula.

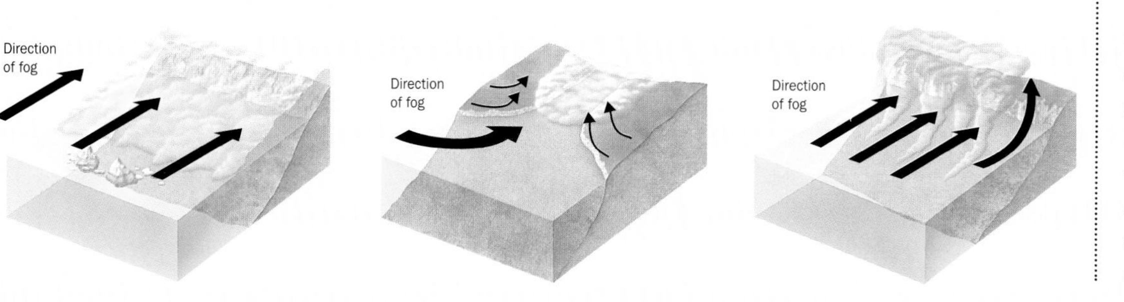

Rolling Banks *Fog forms along coasts: when a cold sea meets warm air from the land – advection (far left); when warm, moist sea air meets a cool landmass – radiation (centre); when cold air passes over a warm sea – sea smoke (left).*

IN DECLINE *The ochre starfish (*Pisaster ochraceus*), a cold-water species, is becoming less common along the California coast as the temperature of the water rises.*

in Monterey waters have been intensely studied since the 1930s, and scientists are beginning to notice changes. The investigation, comparing the animals and plants present in 35 plots studied in 1931 and 1932 with those present now, has identified the first definite link between the steady rise in global temperature this century and changes in marine life.

The researchers divided the marine community into two groups: a southern group, which includes warm-water species that do not live much farther north than Monterey Bay, and a northern group, including species for which Monterey is the southern limit of their distribution. Eight out of nine southern species were significantly more abundant in 1993 than in 1931-2, whereas five out of the eight northern species, including the starfish *Pisaster ochraceus*, were found to be less common. The cold-water species had moved out and the warm-water ones had moved in.

Among the plants, the scientists noticed a dramatic change in the species of seaweed present. During the 1930s a thick blanket of a brown seaweed, very common along the north-west coast of North America, covered the shore, whereas by 1993 this had been replaced by a carpet of red seaweed more characteristic of warmer waters. Eliminating other factors, the researchers concluded that the small rise in temperature had been responsible for this striking ecological change.

Similar but less well studied changes have been reported from other parts of the world. In south-west England a species of sea horse, *Hippocampus ramulosis,* was caught in Plymouth Sound, on the south coast of Devon, in 1990 – the first for 30 years. This 6 in (15 cm) long species is an occasional visitor to the English Channel, but is more normally found farther south, from the Bay of Biscay to west Africa. It lives amongst seaweeds and in sea grass meadows where it anchors itself to the fronds or leaves with the aid of a prehensile tail.

The coral reefs off Jamaica are also showing dramatic changes. Warm water above the critical temperature of 30°C (86°F) causes the tiny algae that live in an intimate symbiotic relationship with the coral animals (polyps) to leave. The algae are responsible for the bright colours of coral: when they have gone, the corals are bleached white. Deprived of its algae, a coral polyp fails to feed properly. If the algae are missing for any length of time, the coral dies.

Why the algae leave is uncertain, though one reason may be that rising temperatures or high levels of ultraviolet radiation put the coral polyps under stress. This could cause them either to withhold nutrients normally taken up by the algae or to produce substances that are toxic to the algae. One thing is certain: the corals tolerate living only in a narrow temperature band, at 25-30°C (77-86°F). If the water temperature rises or falls significantly outside this range, the corals suffer. If the climate change continues, reefs will not be able to follow the rising sea level, and the shoreline will lose its protective barrier and be exposed to severe damage from storms and hurricanes.

CORAL REEF BONANZA

Paradoxically, global warming may prove to be advantageous to some groups of corals. If sea levels rise, slow-growing corals will benefit little, but the fast-growing species found on reef flats will increase rapidly despite rising sea temperatures.

Corals need to be at a depth of between 6½ ft (2 m) and 100 ft (30 m) beneath the surface of the sea for the symbiotic algae living in their tissues to benefit from high enough levels of sunlight. During the past 10 000 years corals have had to contend with a sea level rise of around 3/8-3/4 in (10-20 mm) per year. Many have failed to keep pace and have died; these are nicknamed the 'give-up reefs'. There are also, however, 'keep-up reefs', which generally keep pace with changes in the sea level, and 'catch-up reefs', which lag behind at first but then put on a spurt if the rise in sea level slows.

Experts believe that, owing to a rise in sea levels, reefs of both kinds will grow more rapidly over the next half-century, increasing the amount of fringing reefs on some continental coasts, such as those making up the Great Barrier Reef of northern Australia. This increase in growth may well offset the current destruction of reefs such as the Great Barrier by pollution and global warming.

CORAL BLEACHING *Coral polyps lose their algae when the sea's temperature rises above 30°C (86°F), and the coral turns white.*

When the Sea Turns Wild

The rewards are great for plants and animals that make their homes along the seashore – but so are the risks. High winds can whip the ocean into a fury; more subtly, changes in ocean currents can have effects felt across the entire globe.

Hurricane, typhoon, cyclone, monsoon, tsunami, storm surge – all these words evoke images of destruction. When the sea turns wild, the creatures of the seashore are sure to be in for an uncomfortable time, even though the cause of their discomfort may not be immediately obvious. Indeed, one of the most destructive forces in nature is what appears to be a mild-mannered weather system in the Pacific Ocean. Its influence is so immense, however, that its effects can be felt across the entire planet. It is known as El Niño, meaning 'the Christ child' in Spanish, because it often sets in around Christmas.

Wild under the Sea *A wave slams into a rock face on Malpeno Island, off Colombia's Pacific coast. Bubbles in the turbulence at the point of impact are up to 10 ft (3 m) deep.*

El Niño is not in itself an unusual event. It is well known to weathermen as part of the phenomenon known technically as El Niño-Southern Oscillation (ENSO) which happens regularly every four to seven years. One of its best-known effects is on the ocean currents that bathe the Pacific coast of South America. Normally the icy Humboldt Current flows from the Southern Ocean northwards up the coasts of northern Chile and Peru. In an El Niño year, however, this pattern is disrupted. Warm water from the Equator flows southwards down the South American coastline, displacing the Humboldt Current off Peru and northern Chile. Instead of a nutrient-rich cold-water current flowing northwards along the coast, there is now a barren warm-water current flowing southwards.

What Causes El Niño?

Why does it happen? Many of the workings of El Niño and what causes it remain a mystery even to experts, but certain things are clear. Above all, scientists know that it is accompanied by a reversal of the usual atmospheric patterns. Under normal circumstances, air along the Equator is heated. It rises, fans out to the north and south, cools and sinks, only to be swept back into the tropical belt as the trade winds. Because of the rotation of the Earth, these winds are skewed at the surface of the globe in a roughly east-to-west direction: hence the north-east trades of the Northern Hemisphere and the south-east trades of the Southern.

In the Pacific, one result of this wind flow is a huge build-up of warm water in the western part of the ocean – the sea level around the Philippines is generally about 23 in (60 cm) higher than it is off Panama's Pacific coast. The picture is a complicated one in which the winds in the atmosphere above drive the currents in the oceans beneath, which in their turn affect the patterns in the atmosphere above. The winds blow the waters of the tropical Pacific from east to west, creating the world's warmest expanse of ocean water off the coasts of the Philippines, Indonesia and northern Australia – ocean surface temperatures there are usually above 28°C (82°F). Hot air over the warm ocean rises, creating an area of low atmospheric pressure. The air fans out in the upper atmosphere and then sinks in areas of the central and eastern Pacific such as Tahiti and Easter Island, creating zones of high atmospheric pressure.

During an El Niño event, however, this normal flow is modified. The trade winds break down – for reasons that scientists have not yet worked out. Warm water is no longer driven from east to west; instead, much of the warm water that has built up in the western Pacific starts to surge back towards

Angry Sea *Waves whipped up by strong winds can punch into coastlines with tremendous force, breaking rocks and eroding beaches.*

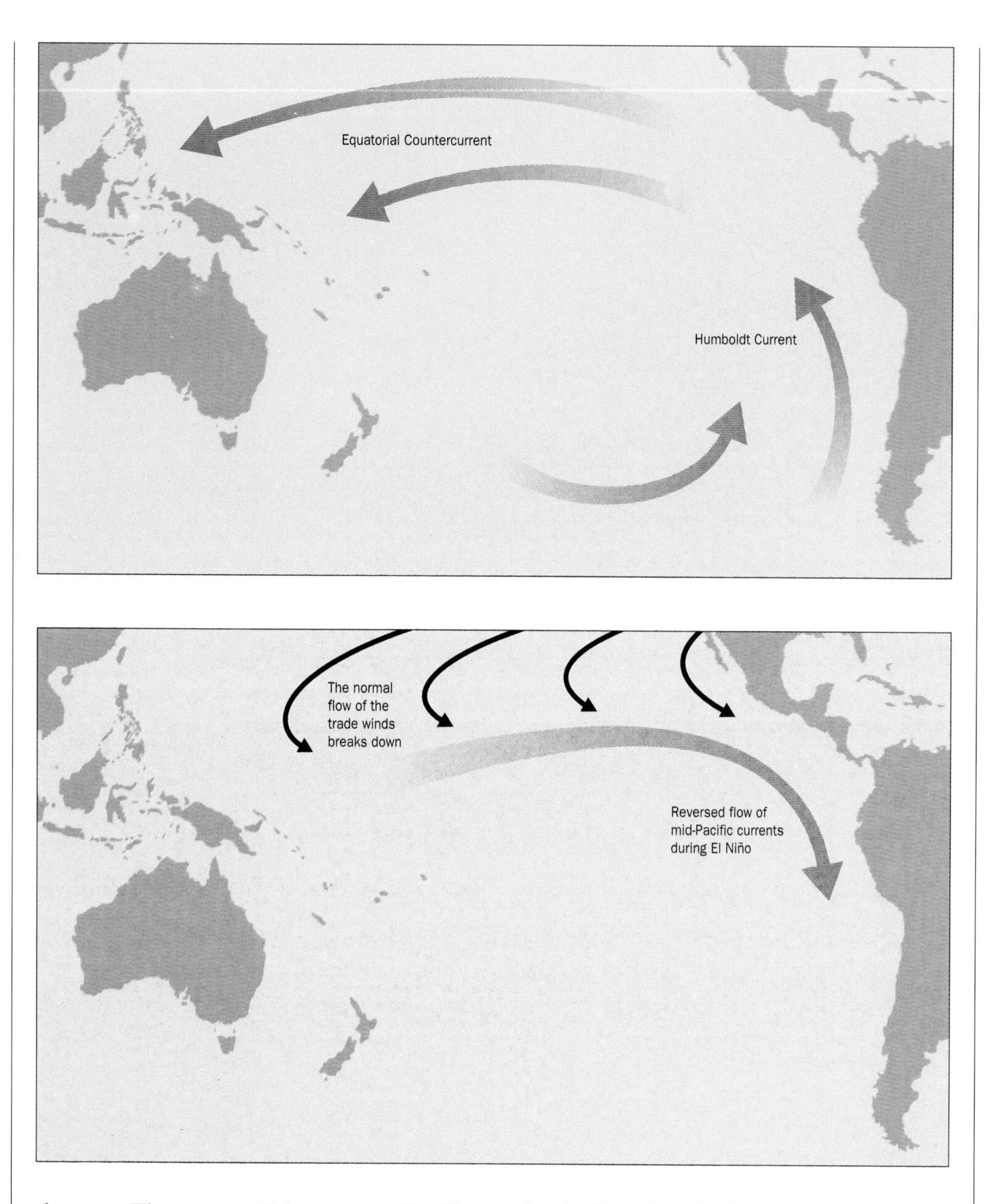

FLOW, COUNTERFLOW *The normal flow of Pacific currents (top) is reversed during El Niño (bottom), with warm equatorial water flowing along the South American coast.*

the east. The waters of the western Pacific are colder than usual, those of the central and eastern Pacific warmer, so that air is now rising over Tahiti and Easter Island and sinking over Indonesia and north Australia. What are normally zones of high pressure thus become low pressure zones, and vice versa.

This whole phenomenon need not be a disaster, and marine and shore-dwelling creatures generally cope with the temporary change of conditions associated with it. Every now and again, however, El Niño strikes with particular force and may result in widespread devastation. This happened in 1982-3. In 1982 an El Niño was slowly developing when, in the spring of that year, the Mexican volcano El Chichon erupted and threw vast quantities of volcanic dust and gases into the atmosphere about 16 miles (25 km) above the surface of the Earth. The explosion was relatively modest, as volcanic eruptions go, but its after-effects were to touch the entire planet, particularly the coasts bordering the Pacific.

After the eruption, the ash and fine dust settled, leaving a mist of sulphuric acid droplets floating in the stratosphere. This acid cloud girdled the Earth and absorbed some of the heat from the Sun, causing a slight drop in global temperature. El Niño was already at work disrupting the normal atmospheric flows in the Pacific, and this drop in temperature intensified the situation. Tahiti recorded the lowest atmospheric pressure for 50 years; by contrast, Darwin in Australia recorded some of the highest pressures this century. In the western Pacific the sea level at Guadalcanal in the Solomon Islands dropped by 6 in (15 cm); in the central Pacific it rose by a similar amount around Fanning and Christmas Islands. A surge of warm water headed for the Galápagos Islands and South America. The sea temperature around the Galápagos rose dramatically: by about 5°C (9°F).

MASSACRE AND SURVIVAL

This was to be no ordinary El Niño. With the usual supplies of moist rising air in the western Pacific cut off, rainfall and other weather patterns were severely disrupted. Drought blighted Australia, where forest fires and red dust storms engulfed vast tracts of land; crops failed in the Philippines. Forest fires in Borneo burnt down an area the size of Switzerland. The drought even reached South Africa and north-eastern Brazil, and the Indian and East African monsoons were interrupted. Powerful typhoons hit Hawaii, and destructive storms sliced through French Polynesia. On the coasts of Ecuador and northern Peru abnormally high rainfall resulted in landslides, burst dams and the accumulation of mosquito-infested water. Among the human population of these regions disease was rife. On the Pacific coast of the United States storms gave rise to extreme high tides, onshore swells and giant waves that eroded coastlines. Throughout the Pacific, wildlife took a pounding. Breeding was severely disrupted, and simple survival became the order of the day.

Unfortunately the 1982 El Niño, which began in May, coincided with the peak of the sea bird breeding period in the North Pacific and continued throughout the year, so that the breeding season in the South

Fires of El Niño *Australian bush fires in 1983, which melted this signpost, were attributed to weather patterns disrupted by El Niño.*

Pacific was affected too. The fish which parent birds would normally feed to their young fled the warm currents and headed for deep, cold waters. An estimated 1 per cent of normal fish stocks was all that remained. Denied nourishment, the birds did the only thing they could: they abandoned their nests and their chicks. All along the Pacific coast of the Americas, from Alaska to Chile, sea birds left their young and flew far out to sea, where the influence of El Niño was less strong. Many did not escape. On the coast of Peru millions of dead birds and fish littered the beaches.

On Anacapa Island, off southern California, brown pelican chicks almost ready to fledge were left to die. On the Farallon Islands, 25 miles (40 km) offshore from San Francisco, the spring upwellings failed to materialise and the seasonal phytoplankton

World Havoc *Abnormal weather conditions, blamed on El Niño, brought a bloom of poisonous algae to the Cape coast of South Africa in 1994, leaving a litter of dead fish, mussels and other shore life. Inset: Peru's band-tailed gull is one of many sea birds that have to adapt periodically to El Niño conditions.*

Here Comes the Rain *Monsoon rainclouds form in the sky off the Yala National Park on the coast of south-west Thailand.*

bloom did not occur. Farther up the food chain, krill were absent and the fish which would normally have fed on them did not appear. Of the 44 000 murres (guillemots) and auklets nesting on the islands, only 500 raised a brood successfully. Few of the 20 000 birds in the breeding colony of Brandt's cormorants bothered even to turn up.

Fur seal mothers on the Galápagos Islands abandoned their pups – indeed, along the South American coast a quarter of all adult fur seals and sea lions died, as did all their pups. Also on the Galápagos Islands, the rare marine iguanas came near to extinction. A drop in salinity on the coast was caused by the run-off of heavy rain. The green seaweeds grazed by the iguanas disappeared from the intertidal zone, the result of changes in salinity and water temperature. These algae were replaced by others that the iguanas could not digest. The higher sea levels also meant that the deeper underwater beds of algae normally grazed by the iguanas were beyond their reach. At the height of the crisis more than half the reptiles were either dead or dangerously underweight.

Ironically, while sea birds and marine iguanas were having a tough time in the Galápagos, other wildlife thrived. Throughout the archipelago more rain fell in two weeks than had fallen in the previous two years. The islands, normally dry and dusty, suddenly burst into life with luscious plant growth. Land birds, unlike their seaside counterparts, laid record clutches. The two species of giant land tortoises also enjoyed a glut of food. Butterflies appeared in profusion. The mosquito and horse-fly populations increased, to everyone's annoyance, and fire ants began to take over territories ten times faster than usual.

Early in 1983 this El Niño terminated; the sea surface temperature in the eastern Pacific dropped, and the trade winds blew strongly once more. Those creatures that had been able to change their diet lived to breed another day. Jackass penguins, for example, had switched to eating squid and shrimps instead of their usual diet of the small fish called anchovetas. Similarly, those that had moved away survived unscathed; they simply missed a year's breeding. Some animals, however, took longer to recover. Fish such as hake and mackerel failed to spawn the following year, and the kelp beds off Oregon, home to the rare sea otter, were reduced considerably, some beds disappearing altogether.

Humans, too, have almost certainly suffered in past centuries from the catastrophic

MONSOON SPAWNING *Triggered by the moon and damp monsoonal winds, red land crabs migrate to the sea to spawn during the rainy season on Christmas Island in the Indian Ocean.*

some parts of the Arabian Sea which are devoid of oxygen, and should the winds bring this water to the surface, fish die in the millions. This happened during the summer south-west monsoon in 1957, when about 19.5 million tons of fish died in an area of about 77 000 sq miles (200 000 km²).

On coasts and islands the rain-bringing south-west monsoon has a vital influence on the cycles of life. On Christmas Island in the Indian Ocean it is one of the triggers that sends huge numbers of land crabs on a pilgrimage to the sea to spawn. In north-western India, in the vast arid plain known as the Rann of Kutch (Kachchh), on the border with Pakistan, the monsoon again plays a key role in a breeding cycle.

With its arrival in June the area is invaded by the sea, to become a knee-deep extension of the Gulf of Kutch. A very rare wild ass lives in this inhospitable place, the Indian wild ass (*Equus hemionus khur*), known locally as the khur. Pregnant mares drop their foals during the monsoon, taking advantage of the small temporary islands that rise from the flooded desert, inaccessible to predators such as dholes (Asiatic wild dogs, *Cuon alpinus*), grey wolves (*Canis lupus*) and tigers (*Panthera tigris*). For food they rely on the salt-tolerant grasses that grow out of the saline sludge of the islands. By the time the waters recede both mother and foal will be better equipped to run away from danger, because khurs can run extremely fast on legs that are longer than those in other species of ass and donkey.

TYPHOONS AND HURRICANES

A far cry from the rhythmic alternation of the monsoons is the sudden destructive fury of a hurricane sweeping across a tropical or semitropical coastal region with winds that may top 155 mph (248 km/h). Its effects can be catastrophic. In August 1992 Hurricane Andrew hit the Florida coast just as sea turtles were nesting. Storm surges flooded nests along a 56 mile (90 km) stretch of shore, destroying most eggs, and beach deposits thrown up by the storm suffocated any hatchlings which survived the flood. This was on top of its devastation to human homes and businesses.

Hurricanes, typhoons and cyclones are all different names for the same thing. Typhoon is the name in the western Pacific, South China Sea and northern Pacific – the word comes from the Cantonese *tai-fung*, 'great wind'. Cyclones happen in the Indian Ocean, hurricanes in the West Indies, Gulf of Mexico and off the southern parts of North America's Atlantic coastline. About a quarter of such storms occur in and around South-east Asia, a seventh in the Caribbean and a tenth in the south-west Pacific and Australian waters.

Essentially, they are massive spiralling systems with winds, clouds and rain storms swirling around a central core, the 'eye' of the hurricane. They start as columns of hot moist air rising from the ocean surface. The column starts to spin, and the chain of events leading to a fully fledged hurricane begins. The spinning air cools as it rises and releases some of its moisture as clouds and raindrops; this process releases latent heat which injects more energy into the system. More hot moist air is sucked in at the bottom and sent spinning up the sides of the central eye, rather like water spinning down a plughole. This leaves a partial vacuum in the centre which draws in air from the upper atmosphere. Unlike the rising air around the sides, which cools as it rises, the descending air in the centre grows hotter. The greater the contrast between the warm air in the centre and the cooling air around the sides, the greater the energy being packed into the system as a whole. The central column of air rises higher and higher, and the winds swirling around it spin faster and faster.

Experts are still not sure why some columns of hot rising air – known as 'seedlings' – develop into hurricanes and others do not. They do, however, know what the right conditions are. Hurricanes depend on a plentiful supply of hot moist air, which means that they form over oceans where the water temperature at the surface is at

WILD ASSES *Khurs, a rare species of wild ass, live in the coastal desert of north-west India which is flooded during the monsoon.*

TYPHOON WATCHING *This spiralling cloud formation, photographed during a Skylab mission, is the vortex of a violent typhoon in the South Pacific.*

least 26°C (about 79°F) – in the Caribbean and Atlantic this applies between June and November, the so-called 'hurricane season'. They also rely on the effect of the Earth's rotation to get them spinning, which means that they cannot form in the latitudes that lie immediately on either side of the Equator where there is not enough rotational force. They never trouble the South Atlantic and Pacific coasts of South America because the cold water currents there keep the ocean temperature too low.

Since the 1970s the US National Hurricane Center has used the Saffir-Simpson scale to classify hurricanes. The range is from a Category 1 hurricane, in which winds reach speeds of 74-95 mph (120-153 km/h), to the relatively rare Category 5, with wind speeds of over 155 mph (250 km/h). As if high winds and torrential rain were not enough, other hazards associated with hurricanes are a rise in the sea level and, more dangerously, a storm surge swamping low-lying shores. Underneath the hurricane, at the sea's surface, two forces are at work: the low pressure at the eye causes a hydrostatic (pressure-related) rise in sea level, while the swirl of spiralling winds causes the storm surge. In deep water the rise may be a few inches only, and on coasts such as island-dotted ones, where there is deep water close to shore, storm surges are insignificant. When the hurricane pushes the sea across an extensive continental shelf less than 300 ft (90 m) deep, however, the surge may reach a height of 16 ft (5 m) or more. One exceptionally high 'wall of water' was recorded on October 7, 1737, when a 39 ft (12 m) storm surge hit the Indian coast near Calcutta. Some 300 000 people died as a result.

Hurricanes are extremely destructive of shoreline plant and animal communities. In August 1980 Hurricane Allen swept across the Caribbean. In Puerto Rico, seaweeds in the intertidal and subtidal zones were ripped from rocks and thrown onto the shore. Entire animal communities, which depend on the algae for food and shelter, disappeared overnight. The same storm ploughed through Discovery Bay in Jamaica and caused extensive damage to the coral reefs. Beds of staghorn coral (*Acropora cervicornis*) were flattened, and the reefs were slow to recover because of predation by coral-eating fish. Brazoria County, Texas, was another, unexpected, victim of Hurricane Allen. Here cattle lands were swamped by the sea and, seven days after the flooding, clouds of a salt-marsh mosquito, *Aedes solicitans*, appeared and started feeding on the blood of cattle. Several cattle were found dead or moribund from loss of blood.

In Puerto Rico, Hurricane Hugo in September 1989 left more attractive traces of its passage. There was a sudden population

FOAM-FLECKED SEA

The entire agitated surface of a hurricane sea is usually covered with a foam caused by breaking waves. When wind speeds are lower than 20 ft/sec (6 m/sec), the sea surface wrinkles but does not break. Surface tension holds it together like a skin. Above this wind speed, the wind transfers more energy than surface tension can cope with into the surface waters, and they need a larger area to carry this energy. This is achieved by breaking up into droplets, each held together by surface tension – hence the foam.

explosion of 15 species of butterflies. Where the winds had blown down trees, caterpillars such as those of *Spodoptera eridania* were able to feed on the leaves of the new vegetation. This was followed, however, by an outbreak of parasitic wasps (ichneumonids) and flies (tachids), for which the caterpillars provided unwitting hosts.

Hurricanes cause widespread damage; the spinning fury of a tornado is far more intense and localised. At sea, tornadoes sometimes form waterspouts, though these tend to be weaker than tornadoes on land. Waterspouts can pose a threat to small boats at sea, and cause damage to shoreline wildlife communities if they move inland. Tornadoes and waterspouts form when a slowly rotating mass of warm air, about 330 ft (100 m) across, is pulled upwards.

Hurricane Force *A wrecked home in Florida (below) is part of the trail of damage left by Hurricane Andrew in 1992. Horizontal rain and bent trees mark the passage of Hurricane Gilbert (right) as it lashes Padre Island, Texas, in 1988.*

TWISTER *Tightly spinning high-speed winds form into a destructive and highly dangerous tornado, photographed in Nevada.*

The vortex is stretched and tightened, so the air spins faster and faster until a ferocious and highly destructive burst of wind results. Waterspouts are also associated with tall cumulus clouds and occur when air rotating near the water surface combines with an updraught to produce a funnel of rotating air that extends from the water surface to the cloud base. A waterspout does not, in fact, draw seawater up to a great height. The sea level at the centre of a waterspout may be raised by only a few feet. The tube in the sky is mainly a rotating cloud of condensed water vapour.

Vortices that form in the sea are whirlpools – bodies of water that spin around either a depression in the sea floor or a cavity in rocks. They can be created when opposing currents and tides meet, and when currents strike offshore rocks. Two famous whirlpools are the Charybdis in the Strait of Messina, between the toe of Italy and Sicily, and the Moskenesstraumen (Moskenes Current), better known as the Maelstrom, to the south of the Lofoten Islands, off the coast of Norway. The Charybdis is generated by the movement of water between the Adriatic Sea and the main body of the Mediterranean; the Maelstrom is a result of strong tidal currents, which flow at a speed of about 7 mph (11 km/h), between the islands of Moskenesoya and Mosken.

TSUNAMIS: ROVING OCEAN WAVES

Tsunamis are sometimes known as tidal waves, but they actually have nothing to do with the tide or tidal currents. They are great surges of water that can travel across vast stretches of ocean and are usually associated with underwater landslides, movements along fault lines, volcanic eruptions and earthquakes. In the case of an underwater landslide, which often occurs at the edge of a continental shelf, the wave is generated when the sea floor drops and water rushes in to take its place. The incoming water fills the void with such power that it pushes up a corresponding bulge of water. This wave then radiates away from the site of the landslide.

Tsunamis can travel undiminished for thousands of miles, moving at extraordinary speeds, exceeding 400 mph (650 km/h) in some instances. In the deep sea they are no more than 3 ft (1 m) high and people on ships at anchor may be quite unaware of the passage of one. Yet seconds later the same people may see an entire shoreline devastated, for when the wave reaches a gently sloping shore it may suddenly rise up to monstrous heights of 50 ft (15 m) or more. One tsunami triggered by an earthquake off the north coast of the Indonesian island of Flores in 1992 was reported to be 82 ft (25 m) high when it hit the shore.

Even this monumental wave seems a bare ripple compared to the giant tsunami which is thought to have hit the east coast of Australia about 105 000 years ago, triggered by a submarine landslide off Hawaii. By the

Vortex *A swirling body of water over a depression on the seabed near St Malo in France forms into a large whirlpool.*

time the wave reached Australia it was 130 ft (40 m) high, but at the islands nearest to Hawaii it is estimated to have reached a staggering 1230 ft (375 m). The northern part of Australia's east coast was protected from its worst effects by the Great Barrier Reef, but the south-east was fronted only by a vulnerable barrier of dunes. All the dunes between 66 ft (20 m) and 164 ft (50 m) high were swept away. Geologists today know this happened because, having examined several headlands along the south-east coast, they have found rocky platforms capped with a layer of fossilised sand dunes dating back about 100 000 years. On platforms set on cliffs less than 130 ft (40 m) high, the dunes are missing and the rocky tops are unweathered. Such was the power of the tsunami that huge boulders weighing 20 tons or more seem to have been plucked from the rocky surface and carried away.

In 1946 a tsunami following an earthquake in Alaska hit the Aleutian Islands with such force that a lighthouse standing 30 ft (9 m) above sea level was destroyed, and several hours later the Hawaiian islands were devastated. The impact of such waves on coastal marine life is overwhelming.

Wall of Water *Volcanic activity in Indonesia sends a tsunami surging through the Sunda Strait. Inset: The Thames Barrier helps to protect London from flooding when winds and tides push the sea against the east coast of England.*

Upstream Surge *The waves of the Severn Bore rush upstream as the incoming tide is funnelled over the flow of the river.*

Plants and animals are smothered by marine sand drawn up from the continental shelf; the survivors are smashed by the force of the wall of water.

Surges, Bores and Red Tides

Large waves can also be generated by winds and tides. Gales are capable of gathering up storm surges over 6-10 ft (2-3 m) high, which can swamp a coast and cause widespread flooding in low-lying areas. Such a surge occurred in the North Sea in 1953; it swamped the east coast of England and the Thames Estuary, flooding London and causing the deaths of more than 300 people.

Another kind of large wave is the tidal bore, formed at high tide when the incoming sea meets the outgoing river and creates a moving wave. Regular tidal bores occur in funnel-shaped river estuaries. As the tide comes in, a wall of water sweeps up the river, the highest forming at the times of spring tides. Several large and many small rivers have bores. The Amazon has the *pororoca*, a 5-8 ft (1.5-2.4 m) wave which sweeps up the river and its tributaries at the time of spring tides. There are also regular bores on the Severn in England, the rivers flowing into the Bay of Fundy in Nova Scotia (where bald eagles, ospreys and sea birds follow the incoming tide), the Hooghly in west Bengal, the Brahmaputra in Bangladesh and the Chien-Tang Kiang, or Fuch'un, a river that empties into the Bay of Hangchow in China.

Another, much more insidious, force in coastal waters is a 'red tide'. This is not a physical phenomenon, but a biological one. It is caused by red algae known as dinoflagellates, which at certain times of the year suddenly multiply ('bloom') to such an extent that the sea turns red. Dinoflagellates produce dangerous poisons, and a bloom can wipe out seashore life. Bivalves, such as oysters and mussels, accumulate toxins from the dinoflagellates when they filter these microorganisms from the water as they feed. People who eat the shellfish become ill. In humans the organisms cause diarrhoeic shellfish poisoning (DSP) and the sometimes fatal paralytic shellfish poisoning.

Dinoflagellates seem to be proliferating, probably because of changes in global temperature and the runoff of phosphates and nitrates from farmland. One recent outbreak was in May 1995, when millions of pilchards were found to be dying from a mysterious toxin which spread 19 miles (30 km) a day across the oceans south of Australia. Dinoflagellates turned out to be responsible.

A similar event took place along the eastern seaboard of the USA in 1993, when the dinoflagellates themselves were carefully studied: an incredible story emerged. The type of dinoflagellate responsible was discovered to have no fewer than 15 stages in its life cycle. At most of the stages it is a form of microscopic amoeba. In the absence of prey it can photosynthesise, and is able to rest up for up to two years as a hard cyst in sediments on the sea floor. But when a shoal of fish swims overhead, the dinoflagellate is transformed into an armoured cell with two whip-like flagella and rises up, releasing poisons as it swims. The poison attacks the fishes' nervous systems, paralysing muscles and destroying red blood cells.

Gasping, some fish hurl themselves onto the land. As the fish suffocate, their skins peel. The dinoflagellate attaches itself to and consumes flecks of shredded tissue. Only then, in the presence of the dying fish, do the miscroscopic killers reproduce.

Fish Massacre *A red tide off the western, Gulf of Mexico, coast of Florida in August 1982 left a trail of dead and dying fish.*

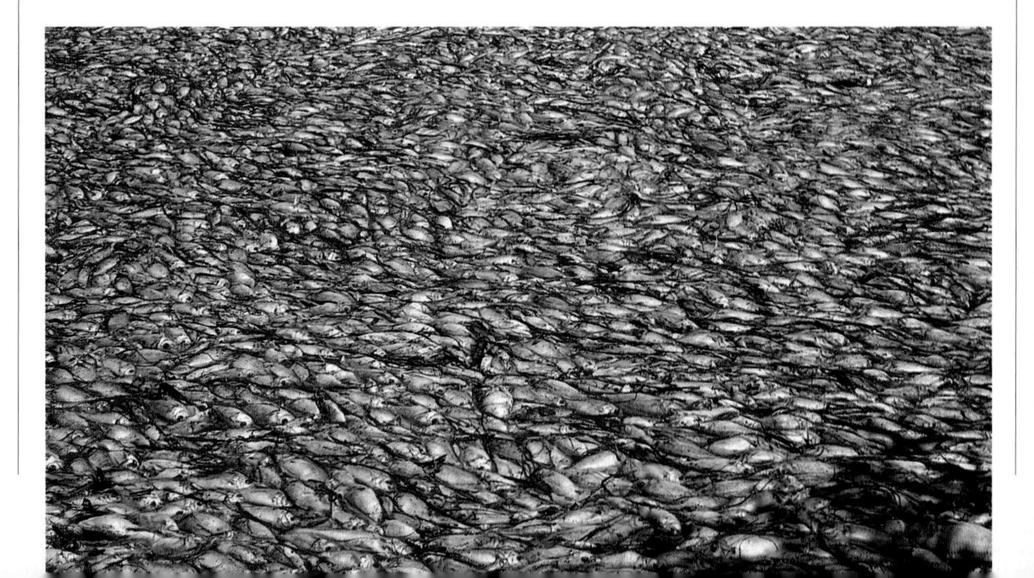

2 Wildlife of the Shore

LIFE IN A POOL *One small rock pool contains several species of seaweeds including wracks.*

SEA BIRDS DOMINATE SEA CLIFFS. ALONG ROCKY SHORES THE RANGE OF LIFE IS CONSIDERABLY MORE EXTENSIVE, FROM CRUSTACEANS TO MOLLUSCS, SEA URCHINS TO GOBY FISH. CRABS AND BURROWING WORMS TAKE THEIR CHANCES ON SANDY COASTS, WHILE ADAPTABILITY — TO ALTERNATE PLUNGINGS IN FRESH AND SEAWATER — IS THE KEY TO THE RICHES OF THE ESTUARY. EVEN THE ZONE WHERE ICE MEETS SEA OFFERS RICH PICKINGS TO THOSE EQUIPPED TO EXPLOIT IT — FROM MICROSCOPIC KRILL TO GHOSTLY LOOKING ICE FISH TO SEALS AND WHALES. ALL CREATURES ARE OPPORTUNISTS, AND THE WILDLIFE OF THE SHORE HAS FOUND AN ABUNDANCE OF OPPORTUNITIES IT CAN EXPLOIT TO ITS ADVANTAGE.

GANNETRY *Gannets nest on Great Saltee off Ireland.*

SEA CLIFFS

Busy, bustling, congested – such is life on the sea cliff. Sea birds in the hundreds, thousands, even millions choose cliffs as safe places to nest and raise their young. Each species has its place in this noisy, competitive environment.

Cliffs make ideal nesting sites for sea birds. Being difficult or impossible for many predators to reach, they are safer than most other land sites. Also, lying right next to the sea, they are close to the nutrient-rich waters of the continental shelf. Congregations of sea birds at cliff sites can be enormous. In the Thule district of north-western Greenland, for example, 10-20 million little auks (dovekies, *Alle alle*) breed in a single colony.

In fact, this extraordinary gathering is just a tiny part of the population explosion that takes place every spring in the high latitudes of the Northern Hemisphere. It is one of nature's most spectacular events. Millions of birds head north to breed, taking advantage of the long days and short nights that enable them to feed more or less round the clock. Swans, geese and waders make for remote tundra sites. Sea birds head for the coast, where most nest on high cliffs in crowded colonies known as sea-bird cities or bazaars.

LIFE IN THE BAZAAR

Most of the Arctic coast of eastern Canada is lined with such metropolises. Over 3.5 million guillemots (murres, *Uria*), 1.5 million fulmars (*Fulmarus glacialis*) and a quarter of a million kittiwakes (*Rissa tridactyla*) arrive in late May and early June. There are an estimated 800 000 nests at just two sites – Digges Island and Cape Wolstenholme, at the northern tip of Quebec's Ungava Peninsula. On the eastern edge of Digges Island the birds nest on a 2½ mile (4 km) cliff face rising 985 ft (300 m) from the sea, while at Cape Wolstenholme an even bigger sea-bird city occupies a 5 mile (8 km) sea cliff.

These sea-cliff colonies are noisy, smelly and look confusing, but there is order here. The birds arrange themselves on the ledges according to their needs, usually on different levels. Shags (*Phalacrocorax aristotelis*) build their nests chiefly near the base of the cliffs, in small caves, on narrow ledges or among boulders – they often return to their nests with waterlogged wings and so would be unable to reach the higher sites. Noisy lines of guillemots (common murres, *Uria aalge*), each with a minute territory, pack onto the narrow ledges of the next storey, while razorbills (*Alca torda*) are scattered about on isolated crevices and behind rocks. Both these birds need a bit of height to get airborne. Farther up still, fulmars and kittiwakes build their seaweed and guano nests on perilously high ledges. Gulls (*Larus*) and terns (*Sterna*) seek open places, where the view is panoramic, and predators such as marauding skuas (jaegers, *Stercorarius*) can be seen approaching; they sometimes nest on plateaus close to the cliffs. Great cormorants

GULLS GALORE *Kittiwakes nest in huge numbers on steep and inaccessible sea cliffs (left). With cliffside nest sites at a premium, territorial disputes (right) are frequent.*

SEAL HAVENS: GIVING BIRTH SAFELY

SEAL ROOKERY ***Inaccessible beaches at the base of high sea cliffs in Cardigan Bay, West Wales, are a favourite breeding ground for Atlantic grey seals.***

Along north temperate coasts such as those in south-west England and Ireland, huge caves on rocky shores backed by high sea cliffs provide safe pupping sites for grey seals (*Halichoerus grypus*). These 'seal caves' have been occupied every breeding season for centuries. They are usually long and narrow, with a beach at the far end and an entrance that is accessible only from the sea. Up to a dozen mothers may be present in a single cave at any one time during the breeding season. The caves are the prime breeding sites, and are first to be chosen when pupping begins. When all of them are occupied, the remaining mother seals are forced to opt for less secure open beaches to give birth.

Common, or harbour, seals (*Phoca vitulina*) tend to choose pebble beaches. For some, however, the only places available to them are rocks. The usual procedure for these seal mothers is to install themselves on a suitable rock slab while it is still covered with water and wait to be left high and dry when the tide goes out. They give birth and then, rather than shuffling across the rocky shore, wait on the slab until the tide comes in again, allowing them to float off. Pups are usually on the move within 40 minutes of their birth. Mother and pup call constantly to each other, each able to recognise the other's voice, so that they do not lose one another, even in stormy seas.

Like the grey seal, the Mediterranean monk seal (*Monachus monachus*) gives birth mainly in sea caves. After a pregnancy of 11 months, a female has a single 44 lb (20 kg) pup. Mother and offspring are vulnerable during the four-month weaning period, and many pups fail to grow up because their mothers are unable to find enough fish to produce the necessary amount of fat-rich milk. Those that do survive remain with their mothers for a further three years. The Mediterranean monk seal is one of the 12 most endangered mammals in the world. Until recently, only 200 of these small seals were thought to have survived in the Aegean and along the North African coast, and a further 100 spread along the coast of Mauritania. Then, in 1993, the known world population more than doubled overnight when a colony of 350 monk seals was found on the coast of war-torn Western Sahara. Ironically, its benefactor has been the war, which has prevented fishing boats from sailing. Fish stocks have built up and the seals have been thriving.

CALIFORNIA BABY ***A female harbour seal and her pup rest on a seaweed-clad rocky shore in Monterey Bay, California.***

they fly, the less food they bring back for the chicks and the less well the latter grow.

At Coats Island, in northern Hudson Bay, Canada, a relatively small colony composed of 25 000 pairs of breeding guillemots feeds its chicks well. At four weeks of age, when they are ready to leave the nest site, they weigh, on average, 8½ oz (240 g). At Digges Island, however, where there are 300 000 pairs in a breeding colony, the chicks weigh only 5½ oz (156 g) at the same age. By this reckoning at least, the Digges Island colony is too big to be as efficient as the smaller Coats Island one.

BIRD TALK

For all cliff-nesting sea birds, sound is a vital means of communication: birds must be able to recognise individual calls in the confusion of a colony. Yet until wildlife sound research began in earnest in the 1960s, biologists thought that visual clues were the main means by which birds found each other. Then they began to realise that although many sea birds have the visual acuity to spot their mates or chicks at a distance of about 650 ft (200 m) or more, sound probably plays an even more important role.

The pioneer of this work was Professor W.H. Thorpe of Cambridge University, in

Early Learning *Guillemot chicks recognise their parents' calls before they hatch.*

England, who visited a gannetry on the Bass Rock in Scotland's Firth of Forth. There he was confronted with a boisterous and very noisy colony of gannets. Even in the gannet's simple squawk Thorpe found that there was enough information for individuals to recognise each other. While recording their calls he also observed their behaviour. He saw that when one of a pair of gannets returned to their craggy, clifftop nest with a crop filled with fish, it flew first to the bottom of the cliff and then hung in the updraught. It rose like a person in a lift, calling all the way. The bird on the nest paid no attention to the calls of other gannets, but as soon as it heard its mate's call it became excited and called back. Thorpe went on to refine his work with gannets and eventually discovered that birds were able to recognise their mates during the first tenth of a second of the call.

Chicks learn their parents' calls even before they hatch from the egg. A guillemot chick, for instance, calls from inside the egg, several days before hatching, ensuring that parents incubate the right eggs. The chick also listens. It must learn its parents' calls; when it has emerged it will respond only to their signals. In a similar fashion, young gulls do not know that they must beg for food unless exposed to their mother's feeding calls. In an experiment, researchers exposed one clutch of gulls' eggs to feeding calls, while another clutch heard no sounds at all. After the chicks had hatched, the researchers found that those that had heard the mother's feeding call pecked at her bill for food. The chicks denied the sound stimulus did not.

Clifftop Snack *A gannet chick on Bass Rock, Scotland, begs from its parent. The parent has returned from sea with a crop filled with fish.*

Sound is important, too, later in life, when a chick becomes a parent itself. Three-quarters of kittiwake pairs team up as lifelong partners, each seeking the other at the start of the new breeding season. The birds cue into each other's distinctive calls. Indeed, a study of kittiwakes in the south-west of England has shown that parent birds not only respond to one another's calls but also tell each other what they intend to do. The 'kuk-kuk-kuk' pre-departure call of the kittiwake tells a partner that the bird is about to leave the nest site. The response of the partner, however, is not always predictable. If the spouse is disapproving, then a great deal of head-tossing follows, and a squabble might ensue, the would-be wanderer put firmly in his or her place. If the mate is comfortable with being left alone for a while, he or she replies with a 'ki-ti-wa-ak' call – the distinctive four-note call which gave rise to the bird's onomatopoeic name – and approval for take-off is given. This call – or one of a number of subtle variations on it that are indistinguishable to the human ear – is given in other contexts, too, such as in fighting, returning to the nest, and when predators are about.

Some birds, such as shearwaters, have local dialects. Researchers have found that a population has to be separated from another for only about six years for even a

NUTRITIONAL EGGS

Gulls feed on the eggs of other birds, even when alternative food such as fish is available. This is because egg proteins contain important amino acids, such as methionine and tryptophan, that are not present in fish proteins. In commercial poultry farming these amino acids have been shown to improve egg production. This suggests that the gulls steal eggs purposefully, and are not just resorting to opportunistic petty pilfering.

human observer to be able to appreciate the difference in dialect. Shearwaters on one island have calls which sound quite different from those on the island next door.

Cliffside Predators

With such dense populations of birds packed onto northern sea cliffs, it is not surprising to find that some predators forsake the rich coastal waters – at least during the summer – and exploit instead the sudden abundance of food that appears, with the onset of nesting, on the cliffs themselves. Many gulls, as well as skuas and ravens, rob nests of eggs, while great black-backed gulls (*Larus marinus*) and peregrines (*Falco peregrinus*) take larger prey.

Great black-backed gulls – the world's largest gulls – avoid breeding in big, dense colonies on ledges, preferring to nest in single pairs or small groups on the tops of remote islands and sea stacks, where they can gain enough lift from air currents to take off. They do, however, visit the sea-bird bazaars for food, which they obtain by predation, scavenging and piracy. At dawn and dusk they intercept Manx shearwaters as these nocturnal sea birds leave or arrive at their nest burrows. They kill them by stabbing them with powerful, stout, sharp bills and shaking them vigorously. The day-active puffins also feature on their menu, caught by the neck either in midair or on the water. The huge gulls skin their victim alive with a deft flick of the bill that literally pulls the body inside out.

Predator and Prey *A puffin returning to its nest burrow has fallen prey to a great skua on a Shetland Island cliff top.*

Unlike the black-backs, peregrines do nest on the sea cliffs. A pair of these magnificent falcons may set up home when about two to three years old, and they will hold the territory in which the nest is sited for the rest of their lives. They guard it extremely fiercely: any passing peregrine that is not a neighbouring territory holder is attacked, while aggressive neighbours are at best tolerated. The nest itself is defended by the larger female peregrine, known as the 'falcon'; the male, or 'tiercel', spends much of his time in spring and summer hunting along the sea cliffs.

Most birds of prey use their ability to fly as a special weapon against creatures on the ground or in the water, but the peregrine has so perfected the art of flight that it hunts expert flyers – other birds – in the air. It does not chase its prey but intercepts it, setting up an ambush in the sky. It waits in the clouds a mile (1.6 km) or so above the ground until, with forward-facing eyes that

Cliff Hunter *A peregrine (left) watches for sea birds which it catches in midair. It brings them back to its nest (below) to feed its chicks.*

ARROWED DIVER *A double-crested cormorant dives after fish off the coast of Florida, USA (opposite). Superbly adapted for swimming but more ungainly on land, a European shag (above) has returned to the shore and hung out its wings to dry.*

give a binocular vision eight times more powerful than ours, it spots a target. It flies out of the sun, folds back its wings, and plummets towards the sea. During this wind-whistling 'stoop', the peregrine reaches speeds of over 100 mph (160 km/h) and, seemingly out of nowhere, strikes down its victim with a single blow. In a flurry of feathers the peregrine's talons break the prey's neck or back and it tumbles towards the shore. The predator then circles and quickly drops down to retrieve its meal. Smaller birds may be snatched, plucked and consumed entirely in midair.

GONE FISHING

The gulls' and peregrines' sea-bird prey are predators themselves, exploiting the abundant shoals of fish that frequent coastal waters. Gannets, and their tropical relatives the boobies, are able to spot shoals of fish tens of feet below the surface, and they can do so from considerable heights. To gain momentum, they adopt the streamlined shape of a torpedo and plunge through the surface in order to snatch their prey. They must also avoid fatal collisions with their flock-mates.

Many other sea birds are adapted to hunt under water. Here these air-breathing creatures are at a handicap compared with the fish and other permanent inhabitants. Another fundamental problem is being able to see the prey, especially when the water is disturbed. Cormorants and shags have a special soft lens which is squeezed into a more bulbous shape that enables them to see better under water. They are accomplished swimmers, with feet set well back on a streamlined body which does not trap air, as this would reduce speed and restrict movement. The bill is sharply hooked and equipped with a hacksaw-like serrated edge for gripping slippery prey. In some

DIVE-BOMBERS *A flock of blue-footed boobies plunge-dives for fish off the Galápagos Islands.*

PUFFINS: THE CLOWNS OF THE AIR

On the Atlantic cliffs of North America and northern Europe a flying circus is staged each summer, performed just as it has been for millions of years. The clowns of the air are puffins, and they attract all manner of predators and pirates, from voracious great black-backed gulls to piratical skuas.

The puffins, drab for the rest of the year while at sea, put on their clown-like faces for the breeding season – between April and August in the Northern Hemisphere. They acquire bright red, yellow and blue plates on their huge, laterally flattened, leaf-shaped bills, and snow-white faces with a blackish triangle around each eye. Tens of thousands may gather close to shore in early spring, ready to occupy their clifftop burrows, changing overnight from solitary sea-goers to colonial land-nesters.

These attractive, comical-looking members of the auk family have serial killers for neighbours – great black-backed gulls. A black-back soars high above, targets a puffin, folds its wings and drops hawk-like out of the sky. But the birds below have their own defence strategy – the presence of other puffins. Great squadrons of puffins circle their nesting sites, the huge numbers of birds forming a 'wheel' that confuses predators and minimises attacks.

The behaviour of the flock amounts to more than just safety in numbers. Split-second timing and precision flying are both vital. The wheel follows a particular preset path around the puffinry, and individuals join it in order to leave or return safely to their nesting burrows. When heading out to sea to feed, an individual will join the carousel and fly a couple of circuits before peeling off and heading for the feeding sites offshore. On its return it rejoins the big wheel, and then drops down to its nest site if no danger is threatening. The gulls, confused by the blizzard of wings, usually fail to home in on their targets, and most of the puffins escape death.

FISHING CLOWN *An Atlantic puffin carries several sand eels in its beak.*

ways they appear to be birds in transition, midway to becoming strictly aquatic creatures in evolutionary development. This can be seen in their wings. Unlike those of a penguin, a cormorant's or shag's wings are not properly adapted to underwater swimming, and are less important in propulsion than the feet. In a cormorant or shag the wing feathers are loose and are not spread with waterproof oil from the bird's preen gland, unlike a penguin's. They therefore become sodden during a dive, which is why cormorants and shags stand on sandbanks or coastal rocks after a dive hanging their wings out to dry.

Many other sea birds of the Northern Hemisphere, such as puffins, guillemots and razorbills, have adapted to an aquatic life to the extent that they appear to be ungainly on land. All have stubby, flipper-like wings and webbed feet set so far down under the tail that the birds stand upright – like penguins that have not yet forgone the option of taking off. For most of their time at sea they are engaged in feeding on fish and shellfish, the guillemot spending between 20 seconds and two minutes at a time under water. Puffins tend to make short dives in quick succession. Indeed, one individual puffin observed diving off the Isle of May in Scotland clocked up 194 dives in 84 minutes. The difference in dive times seems to suggest that the two species are exploiting different resources at different depths, the guillemot diving deeper than the puffin.

THE LONG JOURNEY SOUTH

Diving from the cliffside plays a special role in the lives of guillemot chicks born in the Canadian Arctic when they set out on their first journey – from the guillemots' summer breeding grounds in the Arctic to their wintering grounds hundreds of miles south. The first stage in this extraordinary voyage is simply getting down from the nesting ledge. At about 14-20 days of age, a chick's secondary flight feathers are sufficiently developed for the bird to be able to glide, if a little clumsily.

The parents use a special 'water' call to persuade the chicks to flap-and-fall to the sea below. To reduce the chances of them being eaten by gulls, they are called down at dusk. As one parent calls from the water, the other may fly down with the chick, shadowing it protectively with its body. The male then takes the chick out to sea, the two birds swimming through the night so that in the morning the chick is safely away from marauding gulls and skuas. This is the time when the birds need to be as far away from the coast as possible.

The chicks' primary feathers have not yet grown, so they are unable to fly; instead, they have to swim 620 miles (1000 km) to their traditional wintering grounds off the coast of Newfoundland. The chicks swim together in a large raft, and each one is aided

FIGHTING PIRATES *Two lesser black-backed gulls fight over food or territory on a cliff top on Skomer, West Wales.*

by its father, who helps it to feed during the journey. The mother travels separately, and will not see her mate again until the following year's breeding season begins, when they will meet up again on the same ledge, at the same nest site as last year, ready to start the whole process all over again. Meanwhile, amongst the youngsters making their way south, those that were less well prepared for the journey fail to make their destination. At some point the low survival rate of chicks combined with periodic lack of food begins to limit the size of the guillemot colony.

COASTAL PIRATES

Some species of sea birds make a living out of thievery. They are known technically as 'kleptoparasites'. A flock of gulls is a prime example. Far from observing an orderly 'first come first served' protocol, the flock will indulge in raucous chases as any gull that has managed to purloin a morsel of food is harangued and harassed by the others. If it drops the scrap, another bird will swoop down to pick it up, and the process continues until the food has been swallowed and is out of sight.

Gulls of the same species are as likely to indulge in this noisy ritual as gulls of different species. In mixed flocks of common gulls (*Larus canus*) and the smaller black-headed gulls (*Larus ridibundus*) around the British Isles, the common gulls leave the foraging to their smaller relatives, then try to rob them of their food. In field studies in Ireland it has been found that a single common gull in pursuit of a black-headed gull will obtain food in only 25 per cent of chases, whereas if five birds are involved in the chase the odds are more in the common gulls' favour, with a 75 per cent success rate. The only drawback for the larger gulls is that the actual success rate is divisible by the number of birds in the chase, and so the exercise is curiously ineffective for any individual within the community of common gulls. It is likely that kleptoparasitism is a brief bout of opportunism, and that any scrap of food obtained in this way is supplementing food acquired elsewhere, such as from the nearest town rubbish tip.

LEFT OR RIGHT-BILLED

A black guillemot (*Cepphus grylle*) takes a fish back to its young by carrying it crosswise in its bill. For a long time it was thought that which way round birds carried the fish – some with its head pointing to the left and others with it pointing to the right – was simply a quirk of the individual. But now it is known that the orientation of the fish is related to the wind. The head points to the side that is facing the wind, and this minimises drag.

COASTAL OPPORTUNISTS *Always alert for a meal, gulls wait for the incoming tide to wash in food or stir up the mud.*

Rocky Shores

Creatures jostle for space on hard, rocky surfaces, clinging on as waves and predators try to dislodge them. Below the low-water mark, conditions are easier. The water teems with life, providing feeding and spawning grounds for many species.

The rocky shore forms the front line in the battle between land and sea. The plants and animals living there are pounded and scoured by waves when the tide comes in, and roasted or dehydrated when it goes out. Yet rocky shores contain a rich and varied fauna. The tenacious inhabitants, representing almost every major group of living animals, occupy every available ecological niche.

Shore crabs, swimming crabs, shrimps and prawns represent the crustaceans; sea urchins, spiny starfish and brittle stars are there for the armoured echinoderms (their name derived from the Greek word for hedgehog); the molluscs appear in the form of octopuses, sea slugs and sea snails; and there are fish in plenty – including gobies and blennies, topknots and butterfish, lumpsuckers and eelpout. The group known as coelenterates, characterised by sac-like body cavities, are represented by sea anemones, such as the snakelocks anemone, with its iridescent tentacles that turn a rock pool into a maritime garden, and the aptly named jewel anemone, which turns a pool into a pirate's treasure chest. There are daisy-like colonial sea squirts, moss-like bryozoans, peacock worms that hide their delicate feather-duster tentacles inside 'drinking straws' made from mucus and silt, and jewel-like blue-rayed limpets that burrow into the fronds of brown seaweeds.

Around the rocky shores of Europe's Atlantic coastline, plantlife too is abundant. A swathe of vegetation is ripped away by winter storms each year, only to grow back again vigorously in spring and summer. Growth is so rapid that it rivals regeneration in a tropical rain forest, and a host of animals exploit this period of plenty.

From above the high-tide mark to below the low-tide level, the rocky shore can be divided into recognisable zones, each home to a particular assemblage of plants and animals – each with its own survival problems and its own ways of feeding, fighting and procreating. On rocky North Atlantic coasts the upper shore, lying above the high-tide line except during the highest spring tides, is dominated by brackish pools, a mixture of salt spray and rainwater. This zone is home to creatures such as two marine snails, the small periwinkle (*Littorina neritoides*) and the rough periwinkle (*L. saxatilis*), which have adapted to living most of their lives out of water.

The middle shore is cloaked by rich growths of brown seaweeds or wracks. This

TREASURE POOL *Tiny jewel anemones crowd into a tidal rock pool in New Zealand, their tentacles ready to capture small organisms.*

PACIFIC COAST ***Swathes of green algae and tidal pools occur on a stretch of sandstone coast in central Oregon, in the US.***

tangle of brown algae starts with the broad fronds of *Fucus polycarpus*, followed by the knotted, or egg, wrack (*Ascophyllum nodosum*), with a single row of air bladders down the centre of each frond. Topping it is a covering of the red seaweed *Polysiphonia fastigiata*. Also growing along the middle shore is bladderwrack (*Fucus vesiculosus*), with paired buoyancy bladders that children love to pop when the seaweed is dry.

The zone just above the low-water mark is the lower shore, home to such creatures as the common piddock (*Pholas dactylus*), a mollusc that produces a luminescent blue-green slime – nobody knows why. Below the low-water mark extends the sublittoral zone where, off the rocky shores of north-western North America, luxuriant kelp forests are home to the sea otter.

THE UPPER SHORE

Different kinds of lichen provide some of the best pickings on the upper shore. Here these rocks are often covered with large black patches as if polluted by oil or carpeted with soot; these oily patches are caused by the lichen *Verrucaria maura*, the sooty ones by another lichen, *Lichina confinis*. At night the small and rough periwinkles, which hide in cracks and crevices by day, come out to feed on them.

A little farther down the upper shore is the barnacle zone, home in its upper reaches to the star barnacle (*Chthamalus stellatus*) and in its lower reaches to the fast-growing acorn barnacle (*Balanus balanoides*). A barnacle looks like a mollusc

continued on page 68

HOME FOR LIFE ***The acorn barnacle, with its feather-like legs, lives cemented to a rock.***

Rocky Shore *Distinct zones on a rocky shore can be recognised by the types of seaweed present, as they grow in well-defined bands between the high and low-tide levels. Different species of wrack tend to dominate the higher levels, while kelps grow on the lower parts of the shore.*

1. Sea squirt (*Ciona intestinalis*)
2. Tangle (*Lamonaria digitata*)
3. Murlins (*Alaria esculenta*)
4. Coat-of-mail shell (*Lepidoplurus asellus*)
5. Sea slug (*Elysia viridis*)
6. Red seaweed (*Ceramium rubrum*)
7. Sea lettuce (*Ulva lactuca*)
8. Common starfish (*Asterias rubens*)
9. Dahlia anemone (*Teatia felina*)
10. Red seaweed (*Ceramium rubrum*)
11. Rock urchin (*Paracentrotus lividus*)
12. Common blenny (*Blennius pholis*)
13 & 14. Beadlet anemone (*Actinie equina*)
15. Common shore crab (*Carcinus maenus*)
16. Green sea urchin (*Psammechinus miliaris*)
17. Green seaweed (*Enteromorpha intestinalis*)
18. Knotted wrack (*Ascophylum nodosum*)
19. Edible periwinkle (*Littorina littorea*)
20. Bladderwrack (*Fucus vesiculosus*)
21. Common limpet (*Patella vulgata*)
22. Plumose anemone (*Metridium senile*)
23. Painted topshell (*Calliostoma zizyphinum*)
24. Lichens (*Verrucaria maura, Xanthoria parietina, Bamilina siliquosa*)
25. Rough periwinkle (*Littorina saxatilis*)
26. Channelled wrack (*Pelvetia canalicula*
27. Sea slater (*Ligia oceanica*)
28. Common mussel (*Mytilus edulis*)
29. Acorn barnacle (*Balanus balanoides*)
30. Bladderwrack (*Fucus vesiculosus*)
31. Dog whelk (*Nucella lapillus*)
32. Smooth periwinkle (*Littorina littoralis*)

24
25
17
26
27
18
28
19
20
29
21
13
30
22
31
32
23

Mussel Bed ***A rocky tidal flat on the Pacific coast of the US is packed with a carpet of California mussels. Their 'togetherness' may keep predators at bay.***

(an invertebrate with a soft, unsegmented body, such as a snail or mussel) but is, in fact, a crustacean (a creature like a crab or lobster with a hard carapace) that has adopted a fixed lifestyle. Just like any other crustacean, it starts life as a free-floating larva, but when the time comes for it to mature into the adult form it drifts downwards and cements its head to a rock. The hard protective plates of its 'shell', or carapace, soon form, lest the waves damage the creature's fragile body. Once in place, the adult barnacle remains there permanently. To feed, it extends its legs through a hole in the top of its shell, and tiny hairs on the legs sift food particles out of the water. It has specialised appendages that remove the particles and pass them into its mouth.

There are also real molluscs in this zone, mainly patches of common mussels (*Mytilus edulis*), which are sometimes capped by the seaweed known as purple laver (*Porphyra umbilicus*). It was once thought that the shell-squatting seaweed posed a potential threat to the mussels, causing increased water drag. This, it was reckoned, made them more susceptible to being ripped away during storms. Research in the United States, however, has now demonstrated that the benefits of a seaweed cap may outweigh the disadvantages.

On the Pacific coast of Oregon and California a commonly found mussel, *Mytilus californianus*, often sports a tufted cap of the red seaweed *Endocladia muricata*. During very cold weather many mussels die, the survivors being those with seaweed caps, the weed insulating them from the cold. The researchers also found that in summer the temperature inside shells with 'hats' is much lower than the outside air temperature; protected mussels can survive in conditions where the air temperature rises above 27°C (81°F) – the temperature at which unprotected mussels succumb to the heat. The evaporation of water from the seaweed caps is thought to cool the shell.

Mussels and barnacles alike fall prey to another inhabitant of the upper and middle shore: the large marine snail known as the dog whelk (*Nucella lapillus*). This creature has a rasping, tongue-like strip known as a radula, which it uses to bore neat circular holes in its victims' shells and extract

Seaweed Home *Brown seaweeds in and around a tidal pool (left) may play host to an entire community of animals, including periwinkles (above).*

the soft body tissues within. Some mussels, however, have developed the knack of fighting back. A dog whelk may take an hour or more to inspect a victim and a couple of days to drill through the shell – but it must keep clear of the mussel's foot. Mussels secrete tough byssus threads from glands in their feet, which they normally use to attach themselves to rocks. When a group of them work together, they can also use their threads for defence. The three or four mussels nearest to the predator produce 20 or more byssus threads and attach them to the dog whelk's shell. By retracting the threads, they are able to 'flip' the whelk onto its back.

Only mussels in groups can do this. Solitary ones, though bigger and with thicker shells, are more vulnerable to dog-whelk attacks. To have the best chance of surviving, mussels stick together.

Hanging by a Thread *A common mussel photographed in the act of secreting the byssus threads with which it attaches itself to rocks.*

The Middle Shore

Dog whelks also have a fondness for smaller snails such as the flat periwinkles (*Littorina littoralis*), edible winkles (*Littorina littorea*), thick top shells (*Monodonta lineata*) and purple top shells (*Gibbula umbilicalis*) – all of which graze among the fronds of mature bladderwrack that grow on the middle shore. Also feeding here are common limpets (*Patella vulgata*), which nibble at the middle shore's young seaweed fronds.

Both limpets and mussels are, in their turn, prey to starfish, although limpets are less vulnerable because they can clamp their shells firmly down on the rock. Some starfish species, however, have discovered how to deal with this inconvenience. They unstick their prey by pulling horizontally rather than vertically, so that the limpets are simply slid off. Faced with a marauding starfish, a limpet does what any

New Species, New Phylum

Until 1995 it was thought that all the main groups, or phyla, of animals had been discovered and described; there were 35 in all. But in December of that year a 36th was found. It was a tiny, bottle-shaped creature with a mouth surrounded by cilia (hair-like growths), clinging by an adhesive disc to the bristles on the mouthparts of a Norway lobster. This unique animal was recognised by science as the new species *Symbion pandora*, and, being unlike any other creature known to science, it was placed in a new phylum of its own, Cycliophora.

MICROCLIMATE: SURVIVING IN A ROCK POOL

While global temperature changes may have devastating effects on those coastal plants and animals, such as corals and sponges, that are permanently submerged, the animals of the littoral (intertidal) zones of the shore have quite another climatic regime to endure. In particular they are – and always have been – exposed to much greater extremes of heat. The shallow shore is cool, even icy cold, in winter, but in summer it is never more than warm. By contrast, the temperature of a shallow rock pool in bright sunlight can rise to 40°C (104°F), while rock-dwellers face the danger of dehydration.

How long an organism is exposed to such extremes of temperature will depend on its position on the shore: the closer it is to the high-tide mark, the more time it spends out of water. At neap tides the higher part of a shore could be exposed for several days. Plants and animals are therefore confined to distinct zones, depending on their tolerance not only of extremes of temperature but also of regular dousings in salt water and infrequent bathings in fresh water during storms. Many have adaptations to cope with such abrupt changes.

When exposed, periwinkles (*Littorina*), for example, either crawl down into fronds of damp seaweed or withdraw to the recesses of their shells, which are sealed off by horny lids, known technically as opercula. The limpet *Patella vulgata* clamps down tightly onto a rock before the surface has dried out, trapping moisture inside its thick dome-like shell. Similarly, the acorn barnacle (*Balanus balanoides*) pulls its legs inside and closes its shell plates. The mussel *Mytilus edulis* pinches the two halves of its shell tightly together, while the shore crab *Carcinus maenas* finds a moist, saline microclimate under a stone. The beadlet sea anemone (*Actinia equina*) retracts its tentacles and contracts into a jelly-like blob, and marine worms withdraw deeper into their burrows as the surface of sand or mud dries. Tropical sea snails squirt water over their shells and are cooled by evaporation.

Simple seaweeds such as the red seaweeds, *Porphyra* (including the edible seaweeds known as laver), can lose most of their water – not from the cells themselves, but from the thick cell walls. They may dry to a paper-like consistency, but soon revive again when wetted.

WATER CONSERVATION ***Beadlet sea anemones (right) and limpets (below) at low water.***

self-respecting prey animal would do: it runs away.

Some run faster than others. On the north-western coast of North America there are two species of limpet; one, *Collisella pelta*, lives mainly high on the shore, while the other, *Notoacmea scutum*, is found lower down. The lower-level species shares its living space with the starfish *Pisaster ochraceus*, a specialist in prising off limpets, but it seems to succumb to the starfish's predatory attentions less often than *C. pelta* does. The reason for the difference is not immediately obvious. Both species can 'smell' the starfish, and both have tentacles that can feel its physical presence. *N. scutum's* tentacles, however, are slightly longer than *C. pelta's*, so it has a head start; more importantly, *N. scutum* is able to take off at high speed (high speed, that is, for a limpet). It moves at about 3/8 in (10 mm) per second, which is twice the speed of a pursuing starfish. *C. pelta*, on the other hand, has a top speed of only about 1/8 in (3 mm) per second, and is therefore outrun by its pursuer.

THE GREAT ESCAPE ***A limpet deters the sucker-lined arms of a starfish by exuding a protective slippery mantle and moving away.***

ROCK-POOL ANEMONES

Pools in the middle shore contain several species of sea anemone, including the plumose anemone (*Metridium senile*), which resembles a powder-puff on a stalk, and the ubiquitous beadlet sea anemone (*Actinia equina*). There are two kinds of beadlet: some are a ruby red colour, others are reddish-green or brown. The beadlet is one of the commonest sea anemones of the rocky shore, a position it has achieved even

Rock-pool Raiders *The turnstone has learned how to steal the anemone's dinner.*

though – or perhaps because – it has done away with sex, preferring to reproduce asexually by brooding clones of itself in its body cavity. The clones are released and float to sites nearby if conditions are right, or they may be swept away in currents.

Beadlets fight for the best sites, the places where they are least likely to be dislodged by winter storms. They move by detaching their basal discs and shuffling across the rock surface, and occasionally bump into one another – in which case both anemones are in for a shock. If tentacles touch, a fight breaks out. The level of aggression varies from individual to individual, depending on their colour. Red anemones, which inhabit the upper and middle shore where space is in short supply, are very aggressive. They have densely packed batteries of stinging cells, known as acrorhagi, that show as blue spots in the collar at the base of the tentacles. When confronted with an opponent, the anemone rears up to its full height of 3 in (7.5 cm), inflates the spots and directs the stinging cells at the opponent. The fight seems to be in slow motion, taking five to ten minutes to get going; then the stinging cells are discharged in one of the fastest movements in Nature, completed in three-thousandths of a second. The beadlet on the receiving end withdraws rapidly – 'rapidly', that is, for a sea anemone.

A Sticky End *A blenny is caught by the poisonous tentacles of a dahlia anemone.*

Red beadlet anemones attract the attention of birds, particularly turnstones, stocky little waders that habitually forage on rocky shores. Although, as their name suggests, the birds usually find their prey by turning over stones, some individuals, foraging after the tide has gone out, have discovered that anemones offer rich pickings. Their stomachs are filled with marine worms and other food, and the turnstones have only to peck at the anemones to get to the convenient food package inside.

Rock pools may also be lined with encrusting coralline red seaweeds, and they may hide rock urchins (*Paracentrotus lividus*) and green sea urchins (*Psammechinus miliaris*), which wait for the tide to return before they can leave their crevices and graze the rocks. Serrated wrack (*Fucus serratus*) grows in pools farther down the shore, its flat fronds providing a surface for many small sessile animals – that is, animals that are permanently attached or fixed. Among them are various filter feeders (animals that use a filtering mechanism to obtain nutrients from the water). They include masses

White Spots *Tubes of lime on wrack fronds indicate the work of the tube worm,* Spirorbis.

of the little worm *Spirorbis spirorbis*, which produces tiny white spiral tubes. By attaching themselves to wrack fronds, the worms are able to reach the nutrient and oxygen-rich waters above the rock surface.

The Lower Shore

On the lower shore the green sea slug (*Elysia viridis*) feeds on the finger-like fronds of the green seaweed *Codium tomentosum*. The sea slug does not digest the entire seaweed, but instead breaks into its cells, sucks out the chloroplasts (the microscopic organs that carry out photosynthesis) and sidetracks them into pockets in its own digestive system, where they continue to photosynthesise and provide it with additional energy. Indeed, *Elysia* – described by one observer as 'a leaf that crawls' – obtains about a third of its sugars from the captive chloroplasts. It must keep eating them, however, for the chloroplasts seem to 'run down' when removed from the seaweed.

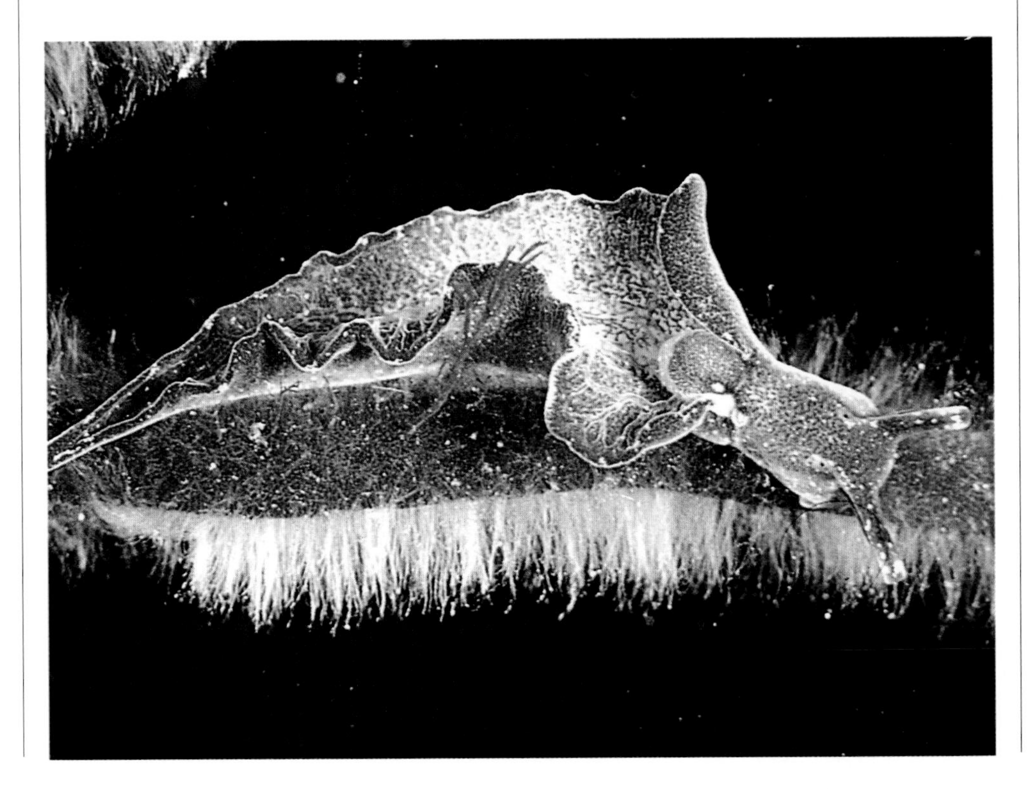

Camouflage *A green-coloured sea slug is well adapted for a life of eating the green seaweed* Codium. *It is almost invisible.*

Where the rocky shore juts into a sandy cove, vegetation is replaced by small reefs consisting of hundreds of tiny tubes constructed by the honeycomb worm (*Sabellaria alveolata*) from grains of sand. A worm builds the first tube by gluing sand grains or shell fragments together. Other worms join it until they have built a 'honeycomb', which may be as much as 2 ft (60 cm) across. At the mouth of each tube is a partition with a hole in the centre. The 2 in (5 cm) worm pokes its head out of the hole and uses tentacles to collect small particles washed in and out by the tide.

Another animal of the lower shore that lives in a tube is the common piddock (*Pholas dactylus*). This remarkable mollusc – a bivalve mollusc like a mussel or oyster – bores its tube through soft rocks, such as chalk, shale and sandstone, using the mechanical action of its shells. Each half of the shell has five rows of fine teeth along its leading edge. The two halves are not joined by a ligament, and the hinge is reduced to a double ball joint. The piddock bores its hole by attaching itself to the rock with its foot and moving its shell valves in a seesaw motion so that they alternately rotate left and right. The teeth at the front grind into the rock, making a hole about 1 in (2.5 cm) in diameter and about 6 in (15 cm) deep. The piddock is a filter feeder, and once established in its tube it extends two siphons – one for inhaling and one for exhaling – out of the tunnel. Internal cilia (microscopic hair-like growths) on the gills generate water currents that draw in oxygenated water and food particles.

ROCK-BORER *The common piddock uses the edges of its shell to bore into soft rocks.*

The seaweeds of the lower shore include the large brown seaweeds known as kelp, such as the strap-like dabberlocks (murlins, *Alaria esculenta*) and oarweed (*Laminaria digitata*). Both of these have long, wavy fronds, each one attached to the rocks by an enormous branching, root-like holdfast.

HONEYCOMB ROCKS *A congregation of honeycomb worms glue particles of sand together to form a reef-like colony that is exposed at low tide.*

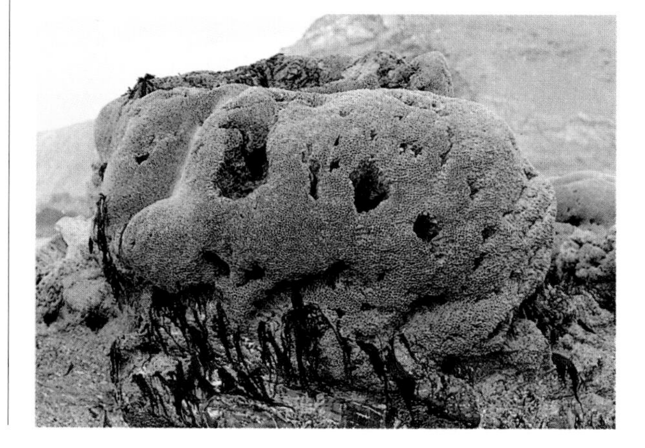

Debris, including organic litter, accumulates among the roots, which can become home to many species of marine creature. Some large holdfasts may have up to 1000 individual animals living among their 'roots' – including hairy crabs, tube worms, sponges, brittle stars, shrimp-like amphipod crustaceans and carnivorous bristle worms, the population changing as first one and then another animal dominates.

The bottom of the kelp fronds and the stipe ('stem') are often grazed by young blue-rayed limpets (*Patina pellucida*), just 1/5 in (5 mm) long. During the summer they feed on the fronds, but in autumn they migrate down the stipe. The fronds are usually badly damaged during the winter storms, and are shed when new growth starts again in the spring. In order to survive the limpets must be close to the base or they will get lost in the sea. Some travel down as far as the holdfast, where a single dominant limpet, which grows to about 3/8 in (1 cm) long, becomes a permanent resident and feeds on the underside of the holdfast, ultimately destroying it and killing the kelp.

BELOW THE WAVES – CALIFORNIA STYLE

Farther down, in the sublittoral zone – the zone below the low-water mark, extending to a depth of about 650 ft (200 m) – the shore is covered by water even during the lowest of low tides. Here waves and currents still influence plants and animals, but life at this level is not exposed to the air or to extremes of temperature at high and low tides. As a consequence it is prolific, and nowhere more so than in the kelp forests off the rocky shores of north-west North America.

Here the kelp fronds grow from the sea floor, held upright by gas-filled bladders that

KELP FEEDERS *A row of blue-rayed limpets feed on a stipe of kelp.*

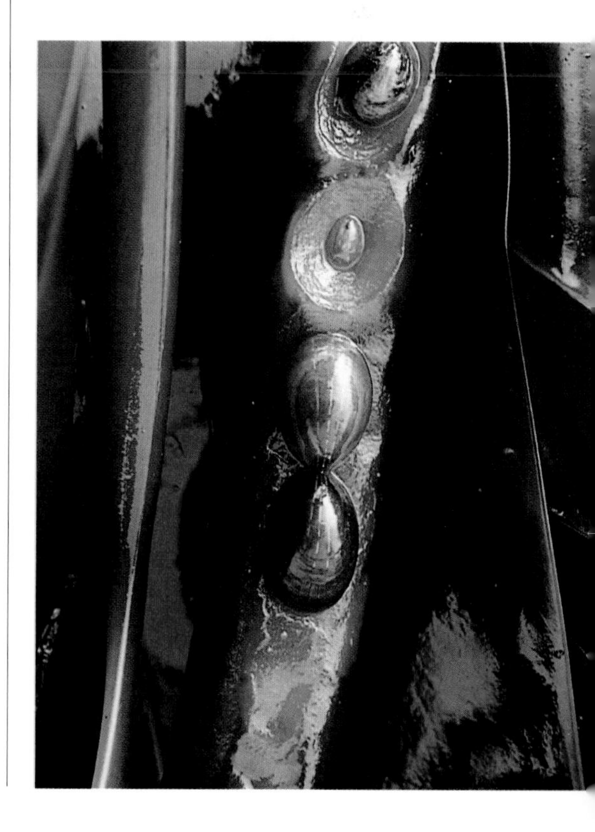

Kelp Forest *Fronds of giant kelp rise from the seabed to the surface. The underwater forest is home to the brightly coloured Garibaldi fish (left), the state fish of California.*

float at the ocean surface about 100 ft (30 m) above. Kelp grows at about 2 ft (60 cm) a day, making it one of the fastest-growing plants on the planet. When it reaches the surface it continues to grow, the fronds eventually reaching over 200 ft (60 m) in length.

One of the most appealing predators of the kelp forest is the sea otter (*Enhydra lutris*). It dives for abalone and sea urchins and cracks them open using improvised tools. Floating on its back, the sea otter balances a stone on its belly, to serve as an anvil on which to smash the shellfish open. Many observers used to regard this as evidence of intelligent behaviour, but subsequent close examination of the development of young otters has shown that this tool-using ability is purely instinctive.

A male youngster observed by Californian researchers in 1985 started the behaviour pattern at about five weeks old, by slapping his chest with his forepaws. At seven weeks he hit an empty abalone shell on his chest with an empty paw, and by nine and a half weeks he pounded two shells together and began to dive. The young otter then dived several times, holding the same piece of rock or shell, just as adults do. Finally, after about five or six months, he was able to crack open a sea urchin on his chest with some degree of success. The youngster had not been influenced by his mother, for he was an orphan in captivity.

In recent years, research on sea otters has revealed many interesting facts. Like any other marine animal, this remarkable mammal must do its best to survive in what is undoubtedly a harsh environment, and this can mean resorting to robbery and violence. Researchers from the Monterey Aquarium and the University of California at Santa Cruz saw young sea otters stealing food from their mothers, and adult males

PROTECTING FUTURE GENERATIONS

The eggs of marine creatures are a nutritious food for predators, so egg-laying animals have adopted all manner of means to make sure that their offspring get a good start in life. These include bulk laying: depositing so many eggs that some are bound to survive. Each summer the inshore waters of Prince William Sound, Alaska, for example, turn a milky white after herring (*Clupea harengus*) congregate there and spawn. Gulls and other birds have a gluttonous field day, but there are so many eggs that the birds are quickly sated, and many eggs survive to develop into young herring.

Dogfish (*Scyliorhinus*) have an alternative strategy for survival. They encase each embryo in a leathery egg-case, equipped with tendrils at each corner to anchor it firmly among the seaweeds in the sublittoral zone. Protected within its tough chamber, the tiny dogfish grows, feeding on a well-provisioned yolk sac, until it has developed sufficiently to survive in the hostile world outside.

GUARD DUTY *A male lumpsucker (right) protects the eggs (above) after they have been fertilised.*

The female lumpsucker (*Cyclopterus lumpus*) deposits her large mass of eggs in shallow waters on the rocky shores of the Atlantic and then departs for deeper water. The male, having shed his sperm and fertilised the eggs, remains with them during the two months they take to develop. He fans the water over them to keep it well oxygenated and guards them from predators until the fish larvae hatch and join the plankton.

The female Mediterranean blenny (*Blennius sphinx*) tests a prospective father's fitness to care for her eggs by giving him a trial sample of eggs to look after. She checks him out a few days later. If the eggs are gone, she does not mate with him. Careful fathers, on the other hand, receive the eggs of many females.

Other such creatures include the pogge or armed bullhead (*Agonus cataphractus*) – a small, 6 in (15 cm) long relative of the minnow found in British waters. It hides its clumps of yellow eggs among the holdfasts of kelp-like brown seaweeds. They are laid in autumn and winter, and take up to 11 months to develop.

The dragon sculpin (*Blepsias cirrhosus*) of the North Pacific lays its eggs in the tissues of sponges. The female fish breaks through the chalky exoskeleton (external 'skeleton') of the sponge *Mycale adhaerens* and deposits clumps of 3-15 eggs in the canals used by the sponge to take in seawater. The fish probably chooses this site because her eggs will be bathed regularly in well-oxygenated seawater. The sponge also secretes antibacterial and antifungal chemicals that keep the eggs free from disease. The eggs remain inside the sponge for about eight months, the young fish eventually swimming free.

MERMAID'S PURSE *A dogfish egg-case, with tendrils that anchor it to seaweeds and rocks, lies on fronds of bladderwrack.*

SEABED SURVIVORS *White sea urchins (above) attack and feed on a larger one. Sea cucumbers (right) void their gut as a defence.*

stealing from any females in their territories. Some males took cubs as hostages until the female handed over her food. Males obtain about a third of their food in this way.

ANTI-PREDATOR STRATEGIES

Predation is the stock-in-trade of some seashore creatures, and their neighbours have had to develop effective defences. Sea urchins that graze the kelp forests encounter numerous predators, including starfish, and the juveniles of one species, *Strongylocentrotus franciscanus*, have found a good way of staying out of trouble. When a slow-moving starfish turns up, the juveniles move as fast as their tube-feet will carry them to hide under the protective umbrella of an adult's spines. Sea cucumbers, sausage-shaped relatives of sea urchins, have their own way of staying alive. When threatened, a sea cucumber forcefully ejects all its internal organs as a decoy, while making good its relatively slow escape. This behaviour may seem rather drastic but, like most other echinoderms, it can regrow the tissues – a process known as 'regeneration' – without suffering any harmful side effects.

For their part, some predators have evolved sophisticated techniques for getting at their prey. The red rock crab (*Cancer productus*) has a penchant for molluscs, especially small molluscs, which are easy to crack open. Large molluscs, however, can present problems. The shell may be too big for the crab to hold in its pincer claws (chelae) and too thick to crush in one squeeze. The red crab overcomes this by holding the bivalve between its body and claw and repeatedly pinching the edge of the shell. It may do this for several days, leaving the shell to feed elsewhere on smaller morsels, but returning regularly to carry out the same procedure.

The crab is utilising a phenomenon more usually associated with crumbling concrete structures, a physical principle known as 'fatigue fracture'. The repeated squeezing at the edge of the bivalve shell enlarges micro-cracks – weaknesses in the shell's structure – until, after about 200 compressions, a fragment eventually breaks away. With the mollusc's defences breached, the crab can force its pincer into the hole and pry the two valves apart. It can then feed at a leisurely pace on the meat inside.

FOREST DWELLERS *The red land crabs of Easter Island go to the shore to spawn. Their eggs are swept out to sea, where the hatchlings develop.*

FLOATING ANVIL *A sea otter (left) cracks open a clam by smashing it against a rock balanced on its belly.*

SHORES OF SAND AND SHINGLE

Worms burrow; crabs and other crustaceans scavenge. Sandy shores are a bountiful home to a range of plants and animals, from palms to salt-tolerant shrubs, from clams to the shy sea-dwelling vegetarian mammal, the dugong.

Sandy shores make up about 75 per cent of the world's ice-free coastlines. They range from no more than a few square yards of sand, tucked away between the rocks of a headland, to great ocean beaches stretching for miles and extending hundreds of feet from land to sea at low tide. These beaches can seem like barren, lifeless places, but in reality they are teeming with life, either hidden away in burrows below the sand or living between the sand grains, too small to be seen by the naked eye.

But, although life abounds, sandy shores do represent an extremely harsh habitat, not only because of the constant pounding of the waves but also because of the instability of the sand itself, which makes it difficult for plant life to take a hold. Intertidal animals are therefore deprived of both a place in which to hide and a resident primary food source. Instead they have to rely on debris from land and sea: dead insects, birds and other carrion from the land, and floating plants and animals, such as plankton and dead or dying jellyfish, from the sea. The creatures of the sandy shore have plenty to adapt to – cycles of deposition

SHINGLE BEACH *A long stretch of beach with sand, shingle and small stones extends along the North Sea coast of England.*

Shore Flower *Sandwort is a pioneering shore plant which can take root even on a shingle beach.*

and erosion of the sand, changing temperatures and, above all, an erratic food supply. They have little room to specialise and are born opportunists.

Top of the Beach

Prickly saltwort (*Salsola kali*), sea rocket (*Cakile maritima*), sea sandwort (*Honkenya peploides*), oraches (*Atriplex*) and sand couch (*Agropyron junceiforme*) are among the salt-tolerant herbs and shrubs that appear in the narrow flotsam and jetsam zone on European beaches, while small holes in the sand just below the strand line – the line that marks the high-water level – betray the presence of at least one form of animal life: sand hoppers (*Talitrus saltator*).

These are a kind of crustacean, 1/2 to 3/4in (13 to 19 cm) long, that emerge at night and scavenge on the debris brought in by the sea. They follow the receding tide some distance down the beach from the strand line, foraging for food as they go, and then, when morning comes, make their way back up the beach to the same area in which they started – here, they dig new burrows.

The ways they navigate differ according to where they live. Mediterranean hoppers have a built-in compass, but those on British beaches backed by sand dunes rely on another stimulus. The British hoppers move towards anything that has a black shape against a light background, such as a sand dune against the sky. There is, however, a fundamental problem. This homing mechanism works well at dawn, when the sand hopper wishes to return home, but it would prevent the creature from going anywhere at dusk, the time when it leaves its burrow to go foraging. The evolutionary answer to this conundrum has been to link the sand hopper's navigational aid to an internal clock: the creature simply switches off its dune-seeking behaviour as darkness falls, enabling it to forage without interruption. When the sun begins to rise, it switches it back on and is guided to its burrowing area.

On tropical shores the clean, sweeping lines of palms dominate. Some palms have seeds that are deliberately cast to the sea to be distributed by the ocean currents. They include the coco de mer (*Lodoicea maldivica*) of the Seychelles Islands, whose large, distinctively shaped fruits, looking like double coconuts, take about ten years to ripen. They float in the currents until they are eventually thrown up on other tropical sandy beaches, where they can germinate and start a new stand of trees.

Sand Living *Sand hoppers live in small burrows on the top of sandy shores and emerge at night to scavenge while the tide is out.*

Real coconuts, the fruit of the coconut palm (*Cocos nucifera*), similarly float from one island to another, buoyed up by their fibrous middle layer (mesocarp). Germination of the stranded seeds is helped when the mesocarp collects rainwater. Where fresh water is scarce, the 'milk' produced by the abundant endosperm (tissue surrounding and nourishing the embryo) helps the seedlings to grow.

On tropical beaches the drift line, the line of debris left on the beach when the tide goes out, is patrolled by a variety of opportunistic crustaceans. These include

CARRIED BY THE SEA *Occasionally a coconut, carried in the currents from one island to the next, may sprout even before landfall.*

mole crabs (*Emirita*) with elongated carapaces (shells) and long, feathery antennae. The mole crab is a fast 'back-burrower'. It tilts forward, using its walking legs to agitate the sand until it becomes semifluid, and then works its body below the surface until only its eyes are visible. It can complete the whole process in about 1½ seconds. When the tide is in, the mole crab sometimes collects food by filtering the water with its antennae and then drawing the antennae through its mouthparts to move the trapped food particles to its mouth.

Ghost crabs (*Ocypode*), with eyes on stalks, are 'side-burrowers'. They dig deep spiralling burrows, to depths of as much as 4 ft (1.2 m), complete with an escape tunnel in the side. They live one to a burrow, each crab indicating that its burrow is occupied by rapping on the sand. It also sends out a strange rasping sound, produced by scraping a series of ridges on the inside of the claw against the 'arm' of the claw; the burrow acts as a resonator that amplifies the sound. The rasping becomes more continuous as the animal becomes more agitated. The sounds are in the 1-2 kHz range and one crab can hear another at about 33 ft (10 m), picking up the airborne vibrations via the cuticle of its legs.

The crabs feed on the strand line, sometimes overturning leaves and catching the sand flies hiding underneath. They eat seaweeds and carrion, including dead sea birds. They are also accomplished predators. A ghost crab can detect the vibrations of a burrowing beach clam (*Donax*) and digs down with one scoop of its walking legs to capture it. Once it has caught its victim it cracks the shell with one of its large claws and consumes the contents. Ghost crabs also eat other crabs. On tidal sand flats they stalk and catch sentinel crabs (*Macrophthalmus*), hauling their struggling victims back to their burrows. Large individuals will even tackle shore crabs, although there is the danger that the predator becomes prey.

Ghost crabs run sideways, often turning through 180° in order to change the leading and trailing legs. They move very fast, scrabbling across the sand at about 5 ft (1.5 m) per second on long, slender legs. They have sensitive eyes, well able to discern a moving object up to 330 ft (100 m) away, and are active mainly at night.

Some creatures ride the waves to get around the shore. They include the beach

Burrower ***The mole crab hides in the sand when the tide is in and becomes active when the tide is out.***

clam, which surfs part-way up the beach with its large foot and siphons extended to take full advantage of the waves. It arrives in the swash zone (the zone covered as waves run up the foreshore), buries itself and uses its siphons to filter feed. After the tide has fallen and risen once more, the clam emerges from the sand and surfs off to a new location. Another wave rider is the whelk *Bullia* which feeds on dead and dying jellyfish, crabs and fish. The whelk is buried in the sand while the tide is out, but when the water returns it enlarges its foot as a float and deliberately uses the waves to reach the nearest fragment of food, which it locates by scent.

In the Waves

In the surf zone of North Atlantic beaches, free-swimming prawns, bass (*Dicentrarchus labrax*) and young flatfish such as plaice (*Pleuronectes platessa*), dab (*Limanda limanda*), sole (*Solea solea*) and turbot (*Scophthalmus maximus*) scour the shore in search of crustaceans, molluscs and worms. These larger animals migrate up and down the beach with the tides, maximising their chances of finding food. Lesser weevers (*Echiichthys vipera*) – which lie below the sand with just their eyes and poisonous dorsal spines showing – and sand gobies (*Pomatoschistus minutus*) also follow the incoming tide. So do sand eels (*Ammodytes*), which feed mainly on other fish. The eels lie buried in the sand or swim in shoals with their heads facing down towards the sand. They, in turn, fall prey to fish such as mackerel and bass and to sea birds such as puffins and terns.

Several species of bird follow the tide in and out. Oystercatchers (*Haematopus ostralegus*) take lugworms, catworms and cockles, while sanderlings (*Calidris alba*) and ringed plovers (*Charadrius hiaticula*) forage for sand hoppers and other small creatures disturbed by the rising waters. In Delaware Bay on North America's eastern seaboard, many sea birds rely on a seasonal beach bonanza for their very survival. In a remarkable conjunction of natural events, millions of migrating waders, on their way north from South America and the southern states of North America to the Arctic to breed, descend on Delaware Bay just as millions of horseshoe crabs are leaving the water to deposit their eggs in the sand. The birds – red knots (*Calidris canutus*), sanderlings, semipalmated sandpipers (*Calidris pusilla*) and turnstones (*Arenaria interpres*) – probe the sand in a feeding orgy that provides the fuel for the rest of their journey north. The birds are in a hurry, for the first to arrive at the breeding ground get the best nesting sites.

The turnstones dig deep. All along the tideline they excavate the crab eggs and

Sand Stalker ***A ghost crab patrols the upper beach (below). The lesser weever's poisonous dorsal fin spines (right) are enough to give the unwary bather an unpleasant sting.***

LIFE AMONGST THE SHINGLE

Seaside Blooms *The sea campion (left) and yellow-horned poppy (above) colonise shingle banks on the coasts of north-west Europe.*

A shingle beach appears to be devoid of life, and in most places it is. The pebbles rub and rotate against each other, preventing plants and animals from gaining a foothold. Thus the intertidal zone is, to a great extent, lifeless. On the top of the beach, however, above the strand line, land plants may grow in the organic debris which accumulates there. On the great shingle beaches of the British Isles, the Atlantic coast of France and the southern shores of the Baltic, isolated clumps of sea campion (*Silene maritima*), yellow-horned poppy (*Glaucium flavum*), sea pea (*Lathyrus crispus*) and sea kale (*Crambe maritima*) occur. On the sheltered lee side of shingle spits the strand line is colonised by shrubby sea blite (*Suaeda fruticosa*), a heather-like plant which prefers well-drained habitats. Nesting on stable shingle banks are oystercatchers (*Haematopus ostralegus*), ringed plovers (*Charadrius hiaticula*) and various species of terns (*Sterna*).

Shingle Camouflage *A tern chick, being fed with a sand eel by a parent, blends in with its stony background.*

squabble for possession of the booty. If a hole is left for a moment other birds will swoop in and jostle for position. Sanderlings often follow the turnstones, fighting for the right to feed at holes. Semipalmated sandpipers share their food, several of them feeding at excavations vacated by the turnstones. Their activity draws the attention of laughing gulls (*Larus atricilla*), which bully their way in and frighten off the less aggressive birds. Even though each egg in itself provides little nourishment, there is such a glut of food overall that migrants can double their weight before they resume their migration. It has been calculated that 50 000 sanderlings alone consume 6 billion eggs, weighing a total of 26½ tons, in just a fortnight. Yet many of the eggs survive, and enough horseshoe crab larvae are washed back out to sea to ensure that the crabs will be back in years to come.

Under the Sand

Filter feeders (creatures that feed by filtering out food particles suspended in the water) and sediment feeders (creatures that feed on deposits on the sea floor) tend to hide under the sand, where they are protected from waves and predators – particularly aerial predators, which scour the relatively even surface of the beach for any sign of disturbance or movement. On European beaches filter feeders include bivalve molluscs, such as cockles – *Cerastoderma* (*Cardium*) and *Acanthocardia*. A cockle is almost spherical in shape and lies just below the sand surface. It has a thick, ribbed shell that resists the force of the waves and the abrasive action of countless sand

Sea Feast *Laughing gulls feed on the eggs of horseshoe crabs which have emerged to spawn in the sands of Delaware Bay on the Atlantic coast of the United States.*

Digging Deep ***An American oystercatcher is able to probe for food with its thick bill. Its diet includes both cockles and lugworms.***

particles. Though out of sight, the animal pushes its siphons above the beach surface and draws in and expels a current of water, from which it is able to extract particles of food in suspension. It is not completely sedentary, being able to move across the sea floor with the help of a mobile 'foot', which it uses like a spring. Often living inside the cockle shell is a diminutive crustacean, the round-bodied, thin-shelled pea crab (*Pinnotheres pisum*), which feeds on the food particles trapped by the cockle, picking them off the gills and even eating the gill tissue itself.

Cockles are hermaphrodites, producing both eggs and sperm which they squirt out via their exhalant siphons. The eggs are fertilised in the seawater and develop into free-floating larvae, known as spat. There is, however, a problem with this method of reproduction: the danger that eggs and sperm will never meet. Another shellfish avoids this difficulty in the most peculiar way. Slipper limpets (*Crepidula fornicata*) live in chains of up to 12 individuals, one on top of the other. The youngest are at the top of the pile, and they are males. As they grow older they pass through a hermaphroditic stage before changing into fully fledged females. The males above fertilise the females below. Several fertilised eggs are placed in a capsule of albumin (egg white) and are incubated under the female.

Sifting Sand

Cockles, together with razorshells, clams, whelks and other filter feeders, are found on exposed shores; more sheltered beaches are populated by many of the sediment feeders, such as bloodworms (glycerid worms) and lugworms. Of these, the most common on

Shellfish Shore ***Storms pile cockles against the shore (left), while the diminutive pea crab (above) hides from the melee inside the shells of a mussel.***

SEX CHANGE CHAIN *Slipper limpets stack in chains, changing sex as they reach the bottom.*

Atlantic beaches is the lugworm (*Arenicola marina*), whose food consists mainly of fragments of organic material, bacteria, the algae that live on sand grains, single-celled animals and other tiny creatures. It has to sift very large amounts of sand in order to obtain these nutrients. Scientists have estimated that only 0.5 per cent of the material the lugworm takes in is edible, so it has to consume up to 1/3 oz (10 g) of sand each day in order to gain the nutrients it needs to stay alive. It lives in a deep U-shaped burrow through which it draws water as well as sand. Waves of muscle contraction ripple from head to tail, moving the water along. The current brings in oxygenated water and loosens the sand in the head shaft, and the increased oxygen levels in the head shaft encourage the growth of the minute algae known as diatoms and other useful food organisms.

A lugworm tends to stay in its burrow for life, although it is able to move, albeit rather clumsily, if evicted. It has a problem with birds. If the sand is wet the worm can continue to feed and defecate, even when the tide is out, but when it sticks its rear end out of the burrow there may be an alert sandpiper or gull waiting, ready to peck at it.

The sand mason worm (*Lanice conchilega*) does not share this indignity, for it emerges only when the tide is high. It also lives in a burrow, but instead of sifting the sediments that fall in, it reaches out of its tube and scours the sand surface. Its tube, constructed of sand particles glued together by mucus, projects above the surface in a tuft, like a miniature tree up to 2 in (5 cm) high. A concentration of sand mason worms gives the appearance of a miniature forest.

The worm feeds when the tide comes in. It climbs up its tube, using minute bristles attached to the front 17 segments of its body, and pushes out a crown of thin, pink, highly mobile tentacles that wipe food

CYCLING SAND: IN THE LUGWORM'S BURROW

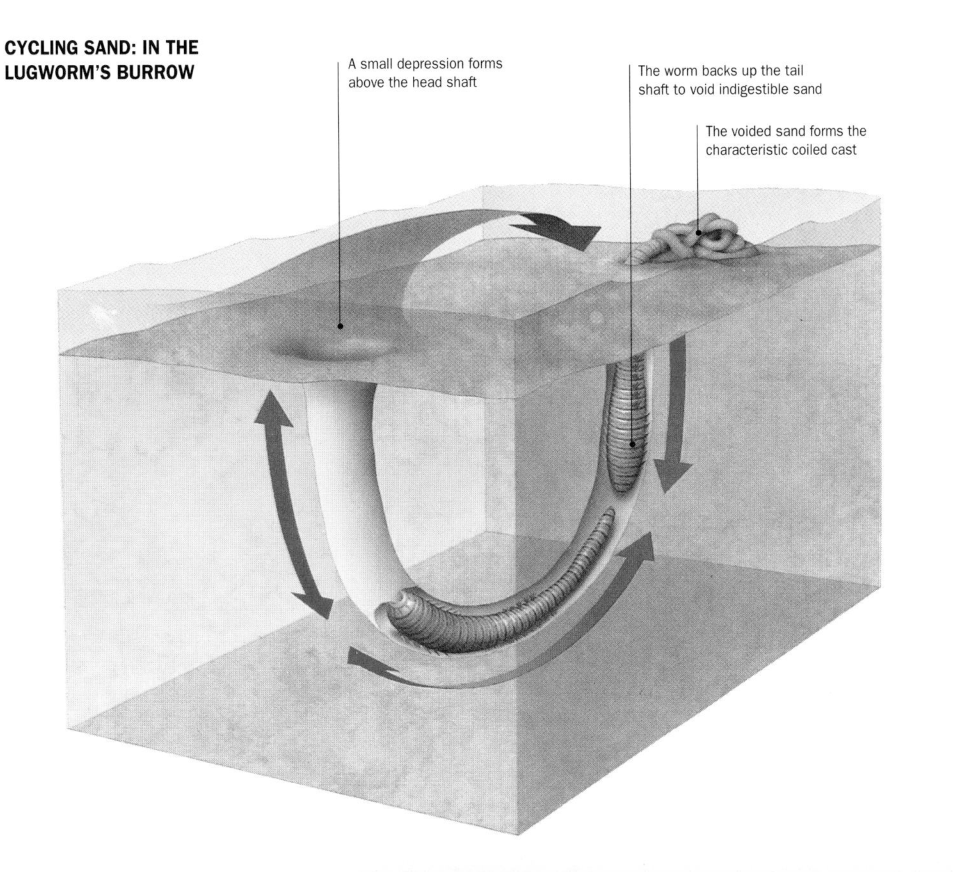

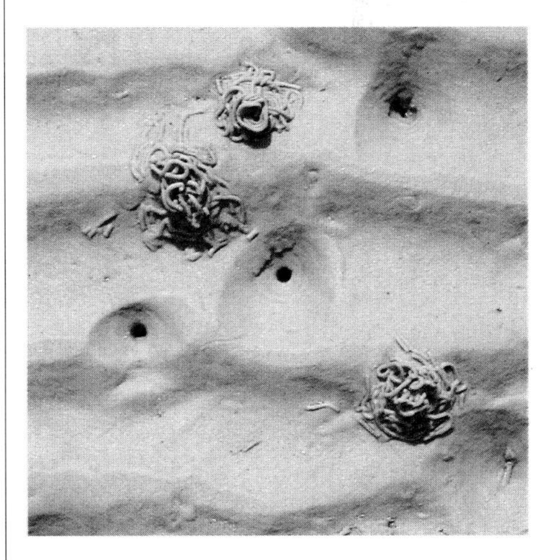

LUGWORM LIFE *Casts on the sand surface (above) indicate the presence of a lugworm (top). The worm lives in a deep U-shaped burrow (left), divided into three sections: the tail shaft, gallery and head shaft. It moves up the head shaft to feed on loose sand.*

Sand Dwellers *Close to the surface of the sand are sand mason worms (left) and the burrowing heart urchin or sea potato (below). Both sift the sand for the tiny organisms that make up their diet.*

particles from the tree-like tufts and also sift the sand around the tube entrance. Small food particles are passed back to the mouth by rows of cilia, while larger morsels are directed to the mouth by the tentacles themselves. Choice sand particles and shell fragments are also sifted out and graded by the 'lips' around the mouth. The larger pieces are taken back to repair or extend the tube, and the smaller ones are used to add to the tufts of sand filaments.

Another deposit feeder of the sandy shore is not a worm but a sea urchin. Unlike other sea urchins, the heart urchin (sea potato, *Echinocardium cordatum*) is not spherical but flattened into a heart shape. It lives up to 6-8 in (15-20 cm) down in a mucus-lined cavity which it constructs and repairs using long tube feet. It can send extra-long tube feet up its respiratory shaft to the surface to pick food particles off the sand. At the front end, modified tube feet surrounding the slit-like mouth collect sand and detritus. The animal feeds mainly on the tiny organisms that adhere to sand grains, holding the sand to its mouth and scraping off any organic material. Unlike other sediment feeders, the heart urchin moves about, not staying in one place for long.

The burrowing worms and small crustaceans and molluscs of the sandy shore fall prey to predatory burrowers such as the burrowing starfish (*Astropecten irregularis*). One shellfish, however, has reversed the roles of predator and prey. This is the 6 in (15 cm) long tun shell (*Tonna zonatum*) found on sandy shores close to Hong Kong. It can detect sea cucumbers – relatives of sea urchins – by smell and devours them whole. After feeding, the tun remains immobile for several days while it digests its food. An even more dangerous predator is the venomous geographer cone shell (*Conus geographicus*), which lives in the sand and rubble of coral reefs. It spears fish with a radula tooth in the form of a poisonous harpoon. Its venom is so powerful that even humans have been killed within hours of being stung.

Undersea Meadows

The sandy shore lacks seaweeds and encrusting animals such as barnacles, for there are no stable, hard rock surfaces to which organisms can attach themselves. But there are some marine plants that grow on sand. These are the eel grasses of the family Zosteraceae, and they form underwater meadows which many marine animals live in or visit in order to feed and breed. Unlike other marine plants, sea grasses are descended from land plants; they have true roots and flower under water. They grow mainly in shallow protected bays or behind reefs and in coastal lagoons.

Among the permanent residents of sea-grass meadows are 35 species of sea horse (*Hippocampus*). Looking like knights in a chess set, sea horses move slowly, propelled by a small quivering dorsal fin on the back. They move in an upright position, but spend much of their time anchored to sea grasses with their tails curled around the stems.

A sea horse's eyes move independently of one another so that it can watch for

Gone Fishing *A cone shell has successfully intercepted a fish and is in the process of devouring it.*

predators while at the same time focusing on a morsel of food (which it sucks in at great speed along its tubular snout). It produces sounds for communication, and can change colour to match its background, which makes it almost invisible. An even more interesting characteristic is the role reversal of male and female sea horses: the male undergoes a complete pregnancy.

Courtship is a lengthy affair. The male sea horse displays the pouch on his abdomen, which brightens in colour and swells, and the female echoes the posture and colour change. Both hold their heads facing slightly away from each other in a 'coy' courtship posture. The male becomes pale and holds open his pouch, while the female approaches. As they face each other, the female inserts her ovipositor into the male's pouch and releases her eggs. She then plays no further role in raising her offspring.

Inside the male's pouch, the eggs are fertilised by his sperm and become embedded in the pouch wall. Here a large number of blood vessels supply oxygen to the developing embryos. The eggs hatch, and the larvae feed on a 'placental' fluid secreted by the male. After 10-15 days, when the salt concentration of the fluid inside the pouch has reached that of the seawater outside, the male goes into 'labour'. For about two days he performs a series of 'hip-thrusts' and squirts out the babies, each one a miniature 1/4-7/16 in (6-11 mm) long replica of an adult sea horse. There may be as many as 14 000 in a single brood, and from the moment of 'birth' they are on their own.

Flowers under the Waves
Sea grasses (above) are among the few flowering plants under the sea. Sea horses mate (below); the male will retain the brood in an enlarged pouch on his abdomen.

A larger and altogether more obvious resident of sea-grass meadows is the dugong (*Dugong dugong*). Living around the fringes of the Red Sea, down the east coast of Africa, along the coasts surrounding the Indian Ocean and in parts of the

GENTLE GIANTS *Dugongs (left) are sea-grass feeders. Glimpses of dugongs – such as this mother and baby (below) – may have given rise to the mermaids of nautical legend.*

south-western Pacific, dugongs follow a more conventional approach to raising offspring than sea horses, although the intimate details of their courtship have only recently been studied.

Dugongs are shy animals. They are mammals that live their entire lives in the sea, but unlike most other sea mammals, which are carnivorous, they are herbivores. They feed on the sea grasses, stripping the foliage or digging up the rhizomes so carefully that not a grain of sand is disturbed. Their only living relatives, the three species of manatee, lead a similar lifestyle in coastal waters in warmer parts of the Atlantic and in some estuaries and rivers.

BEACH VISITORS

Sandy beaches are the temporary homes of other sea mammals, particularly fur seals and sea lions. They come here for one purpose only – to breed. They haul out, sometimes in huge numbers, the males arriving first to establish the right to mate with the females, which follow later. The males of both fur seals and sea lions are fierce fighters, with thick furry manes resembling those of real lions. Those of a similar size come to blows, the mane and blubber around the neck taking the brunt of slashing teeth. Eventually there are winners and losers, and the dominant males gather together groups of females. They are the season's 'beachmasters', and only they court and mate. They must be large and well stocked with blubber, for throughout the breeding season they have

Seal Beach *Fur seals haul out on South Georgia (above). A southern sea lion pup is tossed (left) by a killer whale.*

to fight off sub-adult males intent on mating with their females, and this means that they do not have the chance to feed.

When the females arrive at the beach, their first duty is leftover business from the previous season. Within two days of hauling out, pregnant females give birth, usually to a single pup – the offspring not of a current beachmaster, but of one of the bulls that dominated proceedings the previous year. Most seals and sea lions practise delayed implantation; that is, they postpone the time at which the fertilised egg is implanted in the walls of the womb, thereby delaying the time at which it starts to develop. The actual gestation period is less than a year, but by suspended development the pup is ready to be born not while the mother is at sea but when she arrives back on land.

For the next six days or so the mother remains with her pup and suckles it. Then, eight days after giving birth, she is in heat again and mates with her beachmaster. The bulls become tremendously active during this time, for most of the females give birth – and are therefore ready to mate – at roughly the same time. Fighting breaks out between neighbouring beachmasters, and pups are often crushed in the melee. On the coast of Patagonia in South America, where southern sea lions haul out on windswept beaches, trouble often comes from young males without harems that have not given up trying to mate. These are the juvenile deliquents of the beach. They gang up in groups of about ten and charge the harems, mating with the females in the confusion before the beachmasters can regain control. In what seems to be frustration, some young males capture pups, harassing them and sometimes killing them.

To make matters worse for the pups, another danger awaits them at the water's edge. Killer whales (orcas, *Orcinus orca*) actually leave the sea, riding waves onto the beach in order to snatch young sea lion pups. Wriggling their huge bulk back into the water, the 20-23 ft (6-7 m) long predators then toss the sea lion youngsters into the air using their tails, much like a cat playing with a mouse. Young orcas have lessons from their parents in this remarkable technique, learning their beaching tricks on other beaches.

SECRETS OF THE DEEP

Beaches occasionally reveal some of the sea's profoundest secrets. Animals from the deep are washed ashore dead but sometimes intact, and a few may be previously unknown to science. Some species of beaked whale, such as Andrews' beaked whale (*Mesoplodon bowdoini*), are known only from carcasses that have been washed up on beaches around the world. They have never been seen alive.

3 Sand, Salt and Ice

Seaside Stalker *The common chameleon uses its long sticky tongue to catch its prey.*

Sand dunes and coastal deserts create their own special worlds in the places where land meets sea. The creatures of the long desert strip lining South America's Pacific coast include 'guano' birds, whose droppings are precious as fertilisers, and blood-sucking vampire bats. Elsewhere, in the zone where fresh water meets seawater, burrowing worms help to create the richness of estuary mudflats. Waders busily scour the flats, probing the mud with their bills, and flatfish come in at high tide to join the feast. In polar regions, meanwhile, penguins and seals, huge polar bears and tiny krill, all manage to make their homes in the zone where ice meets sea.

Floating Isle *An iceberg drifts southwards off Greenland.*

Sand Dunes and Desert Shores

Wind-blown sand is constantly shifting, yet it can also be stabilised by plants to create dunes and be invaded by animals. In coastal deserts, sand can be home to an extraordinary collection of animals that rely on fog for their survival.

Natterjack toads (*Bufo calamita*) lie buried in the sand by day, emerging at night to feed on insects. A variety of songbirds, including skylarks (*Alauda arvensis*), reed buntings (*Emberiza schoeniclus*), stonechats (*Saxicola torquata*) and warblers (*Acrocephalus schoenobanus*), live and nest in the dunes, adding a tuneful counterpoint to the harsh mewings of the gulls wheeling overhead. Rabbits (*Oryctolagus cuniculus*), brown rats (*Rattus norvegicus*), field voles (*Microtus agresti*) and wood mice (*Apodemus sylvaticus*) scurry through the scrubby undergrowth. Despite their barren appearance, sand dunes in Europe, as in most other parts of the world, can be rich in all kinds of wildlife.

Dunes are an extension of the beach beyond the high-tide mark. Blown by the wind, sand advances inland, the strength of the wind determining how impressive the resulting dunes are. A minimum wind speed of 10 mph (16 km/h)

CAPE SHORELINE *Sand dunes and drought-resistant, salt-tolerant plants dominate the coast at the southern tip of South Africa.*

PIONEERING PLANTS *A sprig of sea rocket and two stalks of sea lyme grass (left) begin the colonisation of a patch of bare, wind-blown sand. When marram grass establishes itself (above), the dune sand is secured and less likely to blow away.*

is needed to shift sand grains: a Force 10 gale – with wind speeds of up to 63 mph (101 km/h) – moves 200 times more sand than a Force 3 breeze, with wind speeds of up to 12 mph (19 km/h). Where the wind blows hard, coastal dune fields can be almost mountainous. The highest in the world are near Brisbane in eastern Australia, where they reach a height of over 900 ft (275 m).

PLANTS TO BIND THE SAND

Plants as well as wind are vital, for plants and beach debris tend to trap wind-blown sand, forming small mounds that gradually grow into larger dunes. On European shores two coarse grass-like plants – sea couch (*Agropyron junceiforme*) and sea lyme grass (*Elymus arenaria*) – are important dune pioneers. They grow at the top of the beach, but can tolerate occasional soakings in seawater during spring tides, and sprout leaves that trap the sand and roots that bind it. They survive as long as the sand builds up at a rate of less than 12 in (30 cm) per year and the height of the dune is less than 6 ft (1.8 m). Beyond that they cannot keep pace, because rainwater drains away rapidly, leading to a shortage of water and nutrients for the plants' growth; as a result they are swamped by the sand and cannot play any further part in the dune-building process. At this point marram grass (*Ammophila arenaria*) takes over.

Marram grass is at home in sand. It proliferates in dry conditions, helped by the way its leaves curl inwards to reduce water loss from the stomata (tiny pores that allow the passage of gases and water vapour into and out of a plant). When smothered by

YELLOW DUNE *Patches of marram grass and areas of bare, shifting sand characterise mobile or yellow sand dunes nearest the sea.*

MEDITERRANEAN COASTS

Coastal Maquis
Introduced agaves (left) provide cover for creatures like the burrowing mole cricket (right) along the Algarve of Portugal.

The coastal lands of the Mediterranean were once cloaked in forests, the trees reaching right down to the water's edge of a tideless sea, but the introduction of agriculture meant that the trees were felled and the land cleared. Nonetheless, native forests, with small stands of holm oak (*Quercus ilex*), cork oak (*Q. suber*) and Aleppo pine (*Pinus halepensis*), have survived in a few remote places in Spain, Sardinia and the former Yugoslavia, offering a glimpse of what life was like before the arrival of agriculture. Goshawks (*Accipiter gentilis*) hunt for prey – mostly other birds – and cicadas (*Cicadetta montana*) produce their monotonous high-pitched sounds.

The coastal regions that have lost their trees are dominated by thick tangles of low bushes and scrub – a unique habitat known as the maquis, from the Italian *macchia*, 'a thicket'. Conditions are not quite desert-like, but in the summer it is certainly arid. The strawberry tree (*Arbutus unedo*), broom (*Genista cinerea*) and tree heather (*Erica arborea*) are dominant plants, as are myrtle (*Myrtus communis*) and the macchia rockrose (*Cistus albidus*). These two plants survive in the dry winds and hot temperatures of summer on Mediterranean shores by using the plant equivalent of suntan oils, to minimise water loss.

If myrtle leaves are held up to the light, they can be seen to be covered in tiny white dots. Each dot is a gland which produces a heavy aromatic oil. In winter, when water is plentiful, the plant has more than enough moisture bathing its tissues. Water is drawn up through the vascular tissue and evaporates through the stomata during the process of transpiration. In the summer, however, when water is scarce, the plant slows down the transpiration process by secreting a covering of oil over the leaf surfaces.

Other plants of the maquis have different solutions to the summer water shortage. The rockrose produces gums, enough for people to collect the excess for the perfume trade where it is known as 'labdanum'. Prickly juniper (*Juniperus oxycedrus*) has small needle-like leaves from which very little water can evaporate.

Most animals of the maquis hide during the hot days. The digger wasp (*Larra anathema*) burrows into sand. Others, such as long-legged centipedes (*Scutigera coleoptrata*) and scorpions (*Euscorpius flavicaudis*), hide in crevices in the rocks. The maquis is also the home of the Mediterranean chameleon (*Chamaeleo chamaeleo*).

There are parts of the Mediterranean coast where searing winds have reduced the plant cover almost to nothing, such as the maritime steppe of Malta and Gozo. The sea squill (*Urginea maritima*) survives here by having a giant bulb, up to 6 in (15 cm) in diameter, which acts as an underground reservoir.

Parasite and Poison
The rockrose parasite (left) and the sea squill with a bulb from which rat poison is obtained (above), grow on the Mediterranean coast.

DUNE FLOWER *Sea holly blooms on the dunes of the Cabin Hill National Nature Reserve on the coast of north-west England.*

new layers of sand it grows rapidly upwards, keeping pace with the rate of burial. It also grows outwards, binding the surface of the sand with a lattice of roots and rhizomes. An entire dune may be consolidated by a single plant, the hundreds of different shoots growing from the same stock.

Once established, dunes are graded according to their stage of development and distance from the sea. Closest to the shoreline are high 'mobile' or 'yellow' dunes. Vegetation consists of tufts of marram separated by patches of bare sand, so the shape of these dunes is changed regularly by the wind. They reach heights of between 33 ft (10 m) and 115 ft (35 m). The next stage is the semifixed dune, where plant cover is more extensive but still not complete. After that comes the final stage, the 'stable' or 'grey' dune. Its name comes from the presence of sheets of grey-coloured dog lichen (*Peltigera canina*). It tends to be lower than the mobile dune, but with 5 per cent humus (organic debris) it retains more water, contains more nutrients and therefore supports more plant life.

In the embryonic stages of a dune's development, when it is still a mobile dune, the lack of water and nutrients and the temperature of the sand are serious constraints for plant life. Levels of humus are less than 1 per cent, so the sand retains very little moisture, and ground temperatures in summer can exceed 40°C (104°F). Nevertheless other plants, as well as marram, are able to take a hold. Common ragwort (*Senecio jacobaea*), with a long taproot that

SAND-DUNE PHARMACY

In Tudor times, from the late 15th century to the beginning of the 17th century, the edible roots of sea holly (*Erygium maritimum*) were dug from sand dunes and candied. The leaves of another sand-dune plant, hound's-tongue (*Cynoglossum officinale*), were boiled in wine and given as a cure for abdominal pains and diarrhoea, while the roots were ground and mixed with water to form a paste that was believed to alleviate piles.

Maturing Dunes ***Flowers of restharrow (above) are often seen on semifixed dunes, while dog lichen (right) is common on stable dunes.***

reaches down vertically into the sand, flourishes on mobile sand dunes, as does creeping thistle (*Cirsium arvense*), with an extensive rhizome system and divided leaves arranged in a rosette to reduce water loss. Also growing here are sea spurge (*Euphorbia paralias*), with its waxy leaves and long taproot, and sea holly (*Eryngium maritimum*), with waxy leaves and spines to deter browsing animals.

By the time a dune becomes semifixed its plant cover may include sand sedge (*Carex arenaria*), with a long, straight row of shoots following the path of its underground rhizome. The scarlet pimpernel (*Anagallis arvensis*) may also have established itself, along with common centaury (*Centaurium erythraea*) and heath dog violet (*Viola canina*). Hound's-tongue (*Cynoglossum officinale*) and restharrow (*Ononis repens*) may be there, too, a dense covering of hairs on their leaves and stems helping to prevent water loss.

By the time the dune has reached its stable or grey stage, two grasses, red fescue (*Festuca rubra*) and Yorkshire fog (*Holcus lanatus*), have replaced the marram grass. A wide variety of other plants will also have joined the dune community, including orchids, buttercups, thyme and wild asparagus (*Asparagus officinalis*). Another pair are yellow rattle (*Rhinanthus minor*) and eyebrights (*Euphrasia*) which have discovered a novel way to survive the dry and barren conditions: they are both partial parasites, and tap into the roots of grasses for water and minerals. Beach sands often contain high levels of calcium from the fragments of pulverised shells; as a result, yellow feather-moss (*Camptothecium lutescens*) and bee orchid (*Ophrys apifera*), which require dry, lime-rich habitats, are at home on stable dunes. On some European dunes the evening primrose (*Oenothera biennis*) – introduced from North America and now naturalised – thrives. Botanists once believed that its yellow flowers opened in the evening, to be pollinated by moths, but in fact they are open all through the day and are pollinated by any number of insects. When they do open at dusk, it is thought to be connected more with humidity than with preferred pollinators.

Evening Meal ***A hoverfly visits an evening primrose flower which has opened at dusk, triggered by changes in humidity.***

In some places, dune systems assume unusual shapes and forms. For instance, Skagen, at the tip of Jutland in Denmark, has several different formations, including crescent-shaped dunes, whose concave sides face towards the prevailing wind, and barchans, also crescent-shaped but with their convex sides facing towards the wind. Dune pastures in the western Highlands and islands of Scotland form a treeless coastal plain, known by its Gaelic name machair, blanketed by marram, fescues and sedges. At the Banc d'Arguin Nature Reserve to the south of Cap Ferret in Mauritania in north-west Africa, there is an island of dunes that rises no more than 10 ft (3 m) above sea level. Marram and sea couch stabilise the sand, but there are also sea bindweed (*Calystegia soldanella*),

sea spurge (*Euphorbia paralias*) and toadflax (*Linaria thymifolia*).

Where dunes have matured, their peaks or ridges may have a low covering of bushes – known collectively as dune-scrub. All of them have berries that are eaten by birds which then distribute the seeds in their droppings, making sure that the bush cover spreads still farther. Sea buckthorn (*Hippophae rhamnoides*) is a common member of this plant community. Its roots bear tiny swellings (nodules) containing bacteria that can 'fix' nitrogen from the atmosphere, so it can survive in places where the soil is low in nutrients. It can be so successful that it takes over entirely. A dune system invaded by sea buckthorn becomes an impenetrable, spiky thicket.

Flowers of Scotland *Buttercups (above) dominate the dune pastures of the machair of South Uist, Scotland, while elsewhere thick dune-scrub is composed mainly of sea buckthorn (right).*

Dune Animals

Animals as well as plants move into the dunes – the more stable the dunes and the thicker the vegetation, the more animals

there are. A common inhabitant is the banded snail (*Cepaea nemoralis*). It has coloured bands on its shell, usually numbering between one and five, that serve to camouflage it among the plants. Different individuals have different coloured bands – brown, pink and yellow-green. The colour determines whether the individual is seasonally safe or vulnerable. Song thrushes (*Turdus philomelos*) nest among the dunes, and snails are their favourite snack. In

Sand Rabbit *Common rabbits inhabit dunes, their droppings fertilising the ground around their burrows where many non-dune plants thrive.*

Dune Creatures *Banded snails (left) mate on the stalk of a dune plant, while a natterjack toad (right) hides in its hole during the day.*

spring, when growth is green and lush, the yellow-green banded snails are invisible, but the brown ones are easily spotted, and the thrushes tuck into them. In autumn the reverse is true: brown banded snails survive, while yellow-green ones are consumed. Across the year, the population evens out.

Among insects, the larvae of butterflies and moths feed on dune vegetation. The yellow-and-black caterpillars of the cinnabar moth (*Tyria jacobaeae*), for example, have a taste for ragworts. Hunting sand wasps (*Ammophila*) patrol the dunes in search of caterpillar prey, which they take back to their tunnels as provision for their young ones. Hoverflies are conspicuous amongst the flower heads, and solitary bees tunnel into the sand to raise their offspring. One species of leaf-cutter bee, *Megachile maritima*, lives exclusively in coastal areas.

Amphibians are represented by, among others, the natterjack toad. As well as burying itself in the sand, this dune-dwelling specialist deters predators through a toxic white secretion from its skin. In spring natterjack toads deposit their eggs in the small ponds or 'slacks' that accumulate between stable dunes – the hatching natterjack tadpoles are the smallest larvae of any European amphibian. But they are not the only amphibians to use these slacks. Others that breed in them are the common toad (*Bufo bufo*), common frog (*Rana temporaria*) and smooth newt (*Triturus vulgaris*). Mosquito larvae also develop here, amid a profusion of freshwater plants such as water plantain (*Alisma plantago-aquatica*) and broad-leaved pondweed (*Potamogeton natans*).

Reptiles are present on European dunes in the form of two species of lizard. The viviparous lizard (*Lacerta vivipara*) lives on the stable dunes and gives birth directly to 'live' young, whereas the sand lizard (*Lacerta agilis*) inhabits the semimobile dunes and lays eggs. Although they are so similar in many other respects, the two

species keep themselves very much to themselves. Among the dune-dwelling mammals, rabbits mainly eat red fescue. Their droppings at the entrances to their burrows provide a nutrient-rich environment in which many other plants grow, including stinging nettles (*Urtica dioica*), dandelions (*Taraxacum officinale*) and lesser dandelions (*Taraxacum laevigatum*).

With such a diverse animal population, it is not surprising to find many species of bird, including a variety of sea birds that nest in the dunes. Three species are particularly common on European dunes: the herring gull (*Larus argentatus*), the lesser black-backed gull (*Larus fuscus*) and the black-headed gull (*Larus ridibundus*). All these gulls nest in large colonies: the one at South Walney, in north-eastern England, contains about 80 000 herring and lesser black-backed gulls. There are also plenty of large bird predators. Short-eared owls (*Asio flammeus*), kestrels (*Falco tinnunculus*) and, occasionally, peregrines (*Falco peregrinus*) patrol the sand dunes in search of easy

Warning Stripes *The stripes of cinnabar moth caterpillars indicate that they contain plant poisons acquired from the dune plants, such as ragwort and groundsels, on which they feed (right).*

Safe Haven *The skylark brings up its family in a nest on the ground surrounded by dune plants.*

prey, including gull chicks, as do red foxes (*Vulpes vulpes*), stoats (*Mustela erminea*) and weasels (*Mustela nivalis*).

All told, the dune community is diverse and thriving but also vulnerable, living as it does on a substrate of shifting sand. Access to water is a limiting factor. Although rain in temperate latitudes is frequent, the water percolates rapidly through the sand. This means that many of the plants and animals in or on sand dunes, like those living in deserts, must rely mainly on dew for their water supply. In tropical climates this dependence has given rise to some surprising relationships and lifestyles.

Where Desert and Sea Meet

Where upwellings of deep, cold water bring nutrients from the seabed to the surface, even desert coasts can be unusually productive. Nowhere is this more so than on the Pacific coast of South America and the Atlantic coast of southern Africa, the one bathed by the Humboldt Current, the other by the Benguela Current

South America's Pacific coast is dominated by a long, thin strip of desert squeezed between the Andes and the sea. In the northern Chilean stretch is the Atacama Desert – the driest and hottest place on Earth lying adjacent to the largest ocean. Temperatures soar to 50°C (122°F) during the day, yet drop to 0°C (32°F) as night falls. At dusk the explosive sound of gunshots breaks the desert silence; the sudden noise is made, in fact, by rocks cracking and releasing the stresses and strains exerted by the extremes of temperature. Yet there is life even in this truly inhospitable terrain. The grey gull or garuma (*Larus modestus*), the commonest gull along the continent's Pacific shore, feeds, courts and mates along the coast, but is remarkable in its choice of nesting sites. To raise its young, it heads 50 miles (80 km) inland to the desert.

The nests are mere scrapes in the sand arranged in loose colonies. Within a colony they can be anything from about 30 ft (10 m) to 1650 ft (500 m) apart. One of a pair of gulls always remains on the nest, using its fluffed-up feathers and drooping wings to shield the chicks from the sun. Chicks and adults blend in so well with their bleak surroundings that they are almost invisible. As a result adult birds, returning at night from a day's fishing at the coast have to rely on sound to locate their mates. They exchange calls, and after feeding the chicks the pair swap places. Before dawn the other partner flies back to the coast. No other creature, save the odd snake, lizard or scorpion, would venture into such a wilderness, so the nesting birds are relatively safe here. This is why they make the daily journey.

At the coast the gulls join a spectacular horde of other sea birds that feed in the productive inshore waters. The grey gulls feed mainly on a burrowing crab, *Emerita analoga*, while the other birds, such as the local species of tern, the Inca tern (*Larosterna inca*), look for fish. It is here that the so-called 'guano' birds nest and roost in their millions, their nitrate-rich droppings forming the basis of a multimillion-dollar fertiliser industry. There are three main species: the guanay cormorant (*Phalacrocorax bougainvillii*), the Peruvian booby (*Sula variegata*) and the brown pelican (*Pelecanus occidentalis*) – all of them fish-eaters. When the local fisheries collapsed because of overfishing, the guano industry also collapsed. Now, however, the birds have returned and are once more converting the fish into droppings.

The guanay cormorant feeds exclusively on the anchovetta (*Engraulis ringens*), a small fish that swims in huge shoals, feeding in its turn on the animal plankton that migrate each night from the depths to the surface waters. The cormorants fly out from their roosts in long undulating lines, skimming the wave tops until they find a shoal. Then they settle on the surface and dive continuously in pursuit of fish. The Peruvian boobies, meanwhile, also feed on anchovetta, but dive on the shoals from a height of about 60 ft (20 m). They drop

Desert Gull *Grey gulls feed along the Chilean shore by day but return to nests in the Atacama Desert at night.*

FERTILISER BIRDS ***A colony of guanay cormorants in Peru builds its nests on an ever-expanding mound of guano deposited by the birds themselves.***

into the ocean, folding their wings at the last moment before impact. The pelicans likewise dive, but at an angle, not vertically like the boobies. In a great splash the pelicans thrust their heads below the surface and gather fish in their enormous bills. Piratical gulls, intent on stealing the fish from right out of their mouths, scream around their heads until the pelicans regurgitate their hard-won food, which is immediately seized upon by their tormentors.

The cormorants nest on piles of guano that have accumulated on coastal promontories and offshore islands, each bird scraping together a circular crater in which to raise its family. There are about three nests to the square yard (or metre), which means that six adults and six chicks are confined to a relatively small space. The birds will nest only where the dark rocks have been sprayed with the grey-white droppings. Small lizards dash in and out of the nests in search of the feather lice that live on the cormorants.

Sharing the coast are breeding rookeries of South American sea lions (*Otaria flavescens*) and fur seals (*Arctocephalus australis*). The male sea lions from this region are the most lion-like of their kind, with blunt, upturned snub noses and a full mane of coarse hair. The dominant male, or beachmaster, having fought other males and won the right to mate, is surrounded by a bevy of smaller females and pups.

On the coast all manner of unexpected guests arrive on the shore to take advantage of the zone of plenty. An oasis hummingbird (*Rhodopis vesper*) was once seen to forage for marine invertebrates amongst the seaweeds, one of the few times a hummingbird has been seen fishing. Vampire bats (*Desmodus rotundus*) come down to the shore, too. They live in the deep caves of the coast, and in the absence of their usual prey – donkeys, cows and sheep – they turn to seals to obtain their night's supply of blood. Like refugees from a horror movie, they crawl menacingly across the beach and creep up on slumbering seals and sea lions. They take a slither of skin from a flipper, and then, using a tongue equipped with special grooves, lap up the blood that flows from the wound. The bat's saliva contains a substance that prevents the victim's blood from clotting. A creature that is visited too regularly by vampire bats can be weakened to the point of death.

During the daytime Andean condors (*Vultur gryphus*) come down from the mountains in search of seal and sea lion

DENSELY POPULATED ***Each nest mound in a Peruvian booby colony is built just out of bill-pecking distance from its neighbour.***

Nature's Dustmen *Turkey vultures feed on the corpse of a sea lion on a beach in southern Peru.*

Desert Coast *Greater flamingos gather on the Skeleton Coast at the edge of the Namib Desert (right).*

placentas as well as dead animals. Rows of turkey vultures (*Cathartes aura*) hang on updraughts of air, alert for anything that resembles food.

Surviving Through Fog

In south-west Africa another community of desert animals relies on the coastal fogs that form when warm, moist air from the tropical ocean passes over the colder areas of upwelling near the shore. The place is the Namib Desert, a strip of desert running down the length of Namibia's Atlantic coast.

The creatures of the Namib have developed all sorts of ways of collecting the water from fog. Among the 200 species of tenebrionid or 'tok-tokkie' beetle are *Lepidochora discoidalis*, *L. porti* and *L. kahani*, which dig ditches sometimes 1 yd (roughly 1 m) long. Above the ditches are two parallel ridges that collect the water the beetles drink. These remarkable insects retain their water by secreting a waterproof layer of wax on the outside of their hard, protective outer cuticles. This reduces the amount of water that evaporates from the body.

Other animals of the Namib have less sophisticated ways of capturing moisture. Snakes and lizards lick rocks, foliage, their skin and even their own eyes. Dune ants, more usually active during the day, emerge from their nests before dawn and stand motionless, the fog condensing on their bodies. Then they drink the droplets from one another.

The plants, meanwhile, are also adapted to take advantage of coastal fogs. A very common dwarf shrub of the fog zone, *Arthraerua leibnitziae*, exists above ground as green stems. The stems have deep grooves running along them which contain their stomata; these protect the stomata from the wind and thus cut down water loss.

The northernmost stretch of the Namib coastline, just south of Namibia's border with Angola, is known as the Skeleton Coast after the numerous shipwrecks that litter the shore there. It is also the breeding ground for 800 000 Cape fur seals (*Arctocephalus pusillus*) that haul out of the water here each year. The upwellings of cold water offshore ensure that there is usually plenty of food.

The abundance of food means that animals not often associated with the seaside – such as lions, elephants, giraffes, antelope and ostriches – turn up in search of a meal. The giraffes arrive for the droplets of water which accumulate on acacia leaves after a fog. The lions come to the beach to kill seals, which are extremely vulnerable out of the water. They then drag them up to 2 miles (3.2 km) inland, to escape the attention of jackals. The lions also scavenge on seals that are already dead, and on stranded whales. Nowadays, however, fewer and fewer lions survive in the area because local herdsmen have shot most of them.

In the Namib *Creatures adapted to desert life include the darkling beetle (above) and the Namaqua chameleon (right).*

SALT WATER AND FRESH

Estuaries, deltas, salt marshes, mud flats and mangrove swamps have one thing in common: they are shorelines that are dominated by the salt of the sea but are also regularly bathed in fresh water from the land or exposed to the air.

Estuaries, where salt water meets fresh water, offer some of the richest of rich pickings; they also present their own very particular challenge to the animals living in them. The wealth of this environment – abundant in food and relatively safe from the worst ravages of the sea – lies on the estuary floor, where the rise and fall of the tides covers and exposes nutrient-rich bottom sediments, some of which have been washed down by the river, some deposited by the sea itself. The chief hazard of estuary living lies in a principle of physics known as osmosis. This is the tendency for water to move across a semipermeable membrane, such as some animals' skins, from a region of low salt content – say, estuary water at low tide – to a region of high salt content, such as the inside of an animal's body.

For half the day, estuary animals are immersed in fresh water and should, in principle, take in water by osmosis until their bodies burst like

EAST COAST SALT MARSH *A fringe of cordgrasses dominates a salt marsh in North Carolina in the US.*

Homeward Bound *Pacific sockeye salmon (left) and Atlantic salmon (above) spend most of their lives at sea, but have to adapt to freshwater conditions in order to return to their natal rivers to spawn.*

water-filled balloons. For the other half of the day they are covered in seawater, when they should lose water until seriously dehydrated. In fact, many creatures living in this environment are able to change the concentrations of their body fluids in order to keep pace with the changes in the water around them. They do this by taking up or eliminating both water and salts in a controlled way. Fiddler crabs (*Uca*), for example, are able to adjust their internal osmotic concentration in this way, and may have to do so twice a day

Crab Paradise *Fiddler crabs emerge at low tide and scavenge on the mud surface of a salt marsh on the coast of Georgia in the US.*

with the change of tide. The salmon, by contrast, has to cope with the change between fresh and salt water on just two occasions during its lifetime – once when it leaves its river for the sea and several years later when it returns to spawn.

Recycling the Debris

An estuary mud flat is one of nature's great recycling plants. Its foundation is the silt itself, usually rich in organic debris – dead plant and animal materials. Bacteria fasten onto the debris, breaking it down and releasing nutrients that are absorbed by green algae. Living within the mud and feeding on this wealth of nutrients are countless tiny animals, including different kinds of worms, molluscs and crustaceans. Some are suspension feeders which use the mud as a safe home, relying for food on the diatoms (green algae) and other microorganisms that float by in the seawater. Others are deposit (or detritus) feeders that live in burrows and feed on microorganisms in the mud itself or on the surface. Tides come and go, constantly stirring the sediments and giving new impetus to the process of breaking down the organic debris. They also bring further nutrients in clouds of phytoplankton that come floating in from the sea.

The burrowing creatures of the estuary live at different layers and in different ways. Some, such as the bristle worm (*Chaetopterus variopedatus*), live in burrows just below the surface of the mud; others, such as gapers (*Mya*), live as much as 24 in (61 cm) down. The peacock worm (*Sabella penicillus*) manufactures a tube of silt, mucus and sand grains that projects a little above the surface of the sandy mud close to the low-tide level. When water covers the mud flats, the

ESTUARY SHARKS

The bull shark, one of the most dangerous and aggressive of sharks, has a remarkable capacity for surviving in fresh as well as salt water. It is a regular visitor to estuaries and is even found in lakes, such as Lake Nicaragua, far from the sea. There are even stories of hippopotamuses in African rivers up to 300 miles (480 km) from the sea being bitten by sharks.

MUD-FLAT MEALS *The mud flats at Montrose Basin on Scotland's east coast (left) are a feeding place for waders and other shore birds when the tide goes out.*

peacock worm extends a colourfully banded crown of feathery tentacles – hence its name. These collect food and pass it to the mouth. Simple eyes detect the shadows of passing fish and other potential predators, to which the worm responds by retreating rapidly into its tube.

Creatures that burrow deeper rely on siphons to connect them with the nutrient-rich surface. The common otter shell (*Lutraria lutraria*) is one of several species of suspension-feeding bivalve molluscs that live in sandy mud. It burrows to a depth of as much as 16 in (40 cm) and has siphons twice the length of its body that reach the surface when the flats are covered with water at high tide. It is unable to retract them totally inside the shell when disturbed or when the tide is out, so they are covered by a thick, protective leathery coat. The peppery furrow shell (*Scrobicularia plana*), a deposit feeder, lives in a burrow about 6 in (15 cm) below the surface and has siphons up to 8 in (20 cm) long. It uses the inhalant siphon like a vacuum cleaner: radiating from its burrow entrance is a star-shaped series of grooves in the mud where it has 'hoovered' food particles from the surface.

The presence of all these burrowing creatures has a threefold effect on the mud's composition and structure. First, like earthworms on land, they work the mud, churning it and aerating it. A typical population of mud-burrowers will turn over the top 4 in (10 cm) of their habitat in less than five years. This creates perfect conditions for bacterial decomposition, releasing more nutrients for other microorganisms to grow. Secondly, the numerous burrows also increase the surface area where mud and water are in contact, promoting the exchange of chemicals between the two domains and creating more space for bacteria to colonise. Thirdly, the crisscross pattern formed by the mucus-lined burrows strengthens the sediments and makes them harder to erode. It has been estimated that a dense population of the burrowing amphipod *Corophium* triples the strength of the sediment and therefore creates a more stable environment for its inhabitants. It also increases the amount of oxygen and nutrient-rich mud on its patch by an estimated 150 per cent.

MICROSCOPIC LIFE *Tiny diatoms attach to particles of mud or drift in with the tide to form the basic food stock of mud flats.*

TUBE-DWELLING *Peacock worms (left) live in tubes of silt, mucus and sand, and filter floating plankton from the water. The bristle worm (right) lives in a burrow beneath the mud and feeds on detritus.*

MUD WORM *Although a suspension feeder and burrower, the ragworm occasionally leaves its burrow in search of prey.*

Some animals contribute to the turnover of mud not only by burrowing and tunnel-building but also by seeking out dead bodies. The netted dog whelk (*Nassarius reticulatus*) buries itself beneath muddy sand when the tide is out, and emerges as it comes in. Using its siphon it detects the smell of dead and decomposing corpses of crabs and molluscs in the water. When a carcass is found, all the netted dog whelks in the area converge and consume the flesh.

An all-round opportunist is the ragworm (*Nereis diversicolor*), which burrows 16 in (40 cm) or more into the mud. It is both a scavenger and a predator. It uses its eyes and antennae to detect food and has a pair of strong jaws set on a short muscular pharynx (proboscis); it pushes these out to grab the food. It is also a suspension feeder. It secretes a perforated mucous plug at the entrance to its burrow and draws water through the plug by undulating its body in the tube. The ragworm then eats the plug, together with its meal of trapped food particles.

Also scouring the mud surface are hydrobia snails (laver spire shells, *Hydrobia ulvae*). These tiny creatures, less than $^{1}/_{4}$ in (6 mm) long, are among the most abundant animals on the mud flats of European estuaries. They follow the tide out, crawling over the mud and foraging for food. As the top of the mud dries out, they bury themselves. When the tide returns, they float up on mucous rafts and are carried back to the place where they started, and the twice-daily cycle starts all over again.

WADDLERS, WADERS, PROBERS AND PICKERS

The shelduck moves along the shore, waving its bill in a scything movement through the sediment as it searches for hydrobia snails and other invertebrates. It detects the snails with its tongue and a sensitive tip to

TIME AND TIDE *Deprived of feeding sites, oystercatchers and other waders (below) roost at high tide and wait until they can feed, like this shelduck (right), on the exposed mud at low tide.*

Coastal Roost *Bar-tailed godwits roost on tidal mud flats.*

its bill, and eats great quantities of its tiny prey. One shelduck in south-eastern England was found to have 3000 hydrobia shells in its gut. This dependence on mud-flat animals can be a disadvantage, however. In one disastrous winter in the British Isles in 1962-3, when the intertidal zone froze for several months, the shelduck population crashed.

Waders, too, are frequent browsers among mud flats – indeed, many species in northern Europe are entirely reliant on this habitat for winter feeding. Bill length dictates the kind of prey they go after. Short-billed plovers (*Charadrius* and *Pluvialis*) disturb just the surface of the mud, picking off anything that rushes for cover. Birds with medium-length bills, such as the redshank (*Tringa totanus*), greenshank (*Tringa nebularia*), dunlin (*Calidris alpina*), knot (*C. canutus*) and purple sandpiper (*C. maritima*), take shallow burrowers – a single redshank has been known to consume 40 000 *Corophium* amphipods in a day. Curlews (*Numenius arquata*), black-tailed and bar-tailed godwits (*Limosa limosa* and *L. lapponica*) and oystercatchers (*Haematopus ostralegus*) probe with long bills for more deeply buried creatures.

Among the curlews and godwits the shape as well as the length of the bills is important, as is the way they use them. Both curlews and godwits detect prey by sight and touch; they probe the mud deeply only when they have found something to eat. The curlew's bill, however, is longer than the godwit's and down-turned; it uses it to probe gently. The black-tailed godwit's is straight and the bar-tailed godwit's slightly up-curved; both species dive straight down, pivoting round as they push. The curlew's less vigorous technique is thought to be a subtle adaptation to outwit its prey. Burrowing worms sometimes move deeper when they feel the vibrations of something moving directly overhead. The curlew, with its curved bill, is able to probe some distance ahead of its feet, and therefore does not give away its presence. The disadvantage is that the curved bill is less strong, so the godwits have a better chance of success when probing firm sand.

Another popular food source is the lugworm (an inhabitant of mud flats as well as the sandy shore). When it protrudes its tail out of its burrow to eject faeces, the flatfish plaice (*Pleuronectes platessa*) is there to take a bite at it. Remarkably, the worm is only temporarily out of action: it simply has to wait until its tail regrows within a couple of

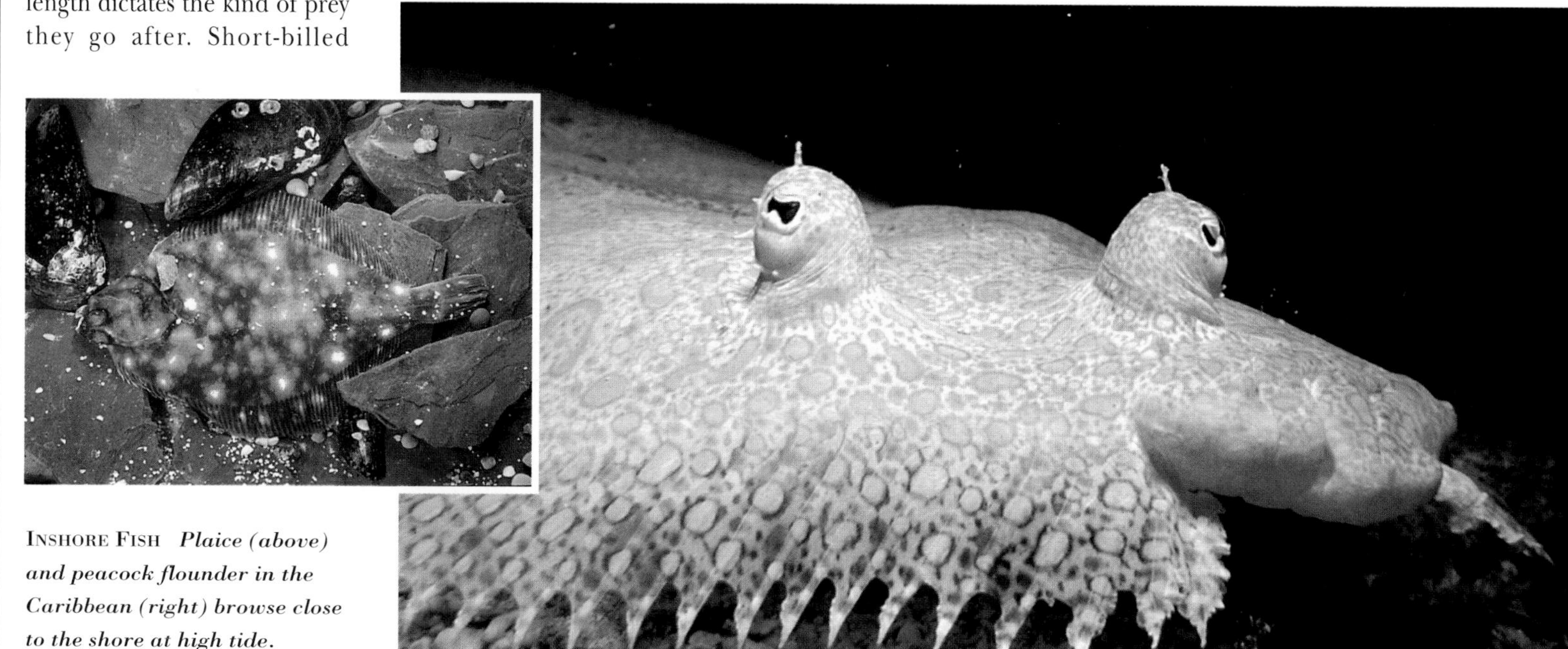

Inshore Fish *Plaice (above) and peacock flounder in the Caribbean (right) browse close to the shore at high tide.*

Salt-marsh Plants *Glasswort and sea purslane (left) and sea lavender (above) grow on European salt marshes.*

months, which means that it will be available to feed other predators time and time again. The process has been likened to a lawn being mowed; the worm is the equivalent of grass, and the predators are the mowers. On the Dutch coast, scientists have discovered that 50 per cent of the diet of young plaice consists of lugworms' tails.

As well as plaice, other flatfish, such as brill (*Scophthalmus rhombus*), dabs (*Limanda limanda*) and flounders (*Platichthys flesus*), arrive at mud flats with the incoming tide. Like all flatfish, the flounder starts out in life in the open ocean as a free-swimming larva, but after about two months it changes its shape and moves inshore. The left eye migrates to the right-hand side of the head, the right side turns a darker colour, and the fish swims with its right side on top. It scours the estuary floor for small crustaceans, and then as it matures starts to prey upon lugworms, ragworms, small shellfish and anything else that might be unfortunate enough to raise its head or tail above the surface of the mud.

European Salt Marsh

Salt marshes are another feature of life where salt water meets fresh water. They form on gently sloping, protected shores – either in sheltered estuaries or behind shingle bars and sand spits – when mud, silt and sand held in suspension in the water are deposited during the brief period of slack water at high tide. Green seaweeds such as sea lettuce (*Ulva lactuca*) and gutweed (*Enteromorpha intestinalis*) bind the mud, and small hummocks separated by shallow channels gradually build up. In some places, eelgrasses (*Zostera*) play the same binding role. A salt marsh takes 150 years or more to reach maturity, at which stage it, unlike a mud flat, is no longer covered by the sea except during spring tides.

The plantlife in salt marshes typically consists of grasses and shrubby plants which must be able to cope with salt water. Many salt-marsh plants have succulent leaves, a feature usually associated with deserts. Although the plants are surrounded by water, it is saline, and they have difficulty taking it up. It is as if they were living in a desert.

Covered in grey silt, the green swards of salt marshes along North Atlantic coasts look uninviting, but there are many plants fighting

Essex Salt Marsh *Cordgrasses grow on a salt marsh on England's east coast.*

SEA OF PURSLANE ***Islands of sea purslane (left and below) are characteristic of a mature salt marsh in Poole Harbour, Dorset, England.***

for a space and, like the seaweeds on the rocky shore, they occupy distinct zones, depending on their height above sea level, the stage of development of the marsh and the length of their immersion in seawater.

The first zone is populated by the green seaweeds and eelgrasses that initially bind the mud, silt and sand. Next, in zone two, come various pioneering plants including rice grasses (*Spartina*) and salt-marsh grasses (*Puccinellia*), which strengthen the mounds while the channels between them are scoured by the tide. They can germinate only if the seeds, washed ashore in the tide, are continuously exposed to the air for at least 50 hours. This patchy cover may also include glassworts (*Salicornia*). The third zone is formed as additional silt is deposited and is stabilised by salt-marsh grass. The fourth zone forms when thrift (*Armeria maritima*), sea-blite (*Suaeda maritima*) and sea aster (*Aster tripolium*), a relative of the garden Michaelmas daisy, establish themselves.

Zones four, five and six of a mature salt marsh are not distinct but are more of a patchwork of vegetation. At the lower levels, salt-mud rush (*Juncus gerardii*) and red fescue (*Festuca rubra*) dominate, while at the higher levels of the salt marsh is a zone of sea rush (*Juncus maritima*). Sea lavenders (*Limonium*), sea purslane (*Halimone portulacoides*) and sea plantain (*Plantago maritima*) are also present. Eventually plants grow across the creeks, obstructing the tidal flow and forming saline pools in which very little grows at all. At high spring tides the marsh is usually covered by seawater.

The salt marsh is invaded by both land animals and the creatures of the shoreline. Land snails and seashore crabs inhabit them, as do migratory ducks and geese, which feed on the marshland plants during the winter. Wigeon (*Anas penelope*) and Brent geese (*Branta bernicla*) graze the eelgrasses, gutweeds and other plants of the lower salt marsh, while greylag geese (*Anser anser*) and Canada geese (*Branta canadensis*) feed at the higher levels. The plants are eaten by smaller creatures, too. The stems of scurvy grasses (*Cochleria*), which grow in the upper zones of the salt marsh, are pierced by the hypodermic mouthparts of the aphid *Lipaphis cochleariae*, while the stems and leaves are devoured by the beetle *Phaedon cochleariae* and its larvae.

If the tide overwhelms the marsh, the birds simply fly away. Other animals have to adopt different survival strategies. Among them is the staphylinid or rove beetle (*Bledius spectabilis*), which has no option but to survive regular swampings. The female deposits her eggs and raises her offspring in a burrow which is covered by the sea to a depth of 3 ft (1 m) or more twice each day. She is an air-breather, yet she does not drown. Her burrow is shaped like a wine bottle, about nearly $2^1/_2$ in (6 cm) long with a chamber about 2 in (5 cm) in diameter, and a neck $^1/_{16}$ in (2 mm) across, which keeps out the sea. The size of the neck is critical. If it were any wider than it is, the

MARSH GOOSE ***Greylag geese are common visitors, feeding on plants growing on the higher levels in the salt marsh.***

MANGROVE FOREST *The trunks, branches and aerial roots of black or grey mangroves (above) provide slipways for mudskippers (left) to emerge from the water.*

burrow would flood, but $^{1}/_{16}$ in (2 mm) is just right for the water to enter the neck and be held by a stable meniscus – a curved water surface, produced by surface tension. The female beetle, however, does not take any chances. About 15 minutes after the tide starts to rise, she takes silt from the burrow wall and, using her mandibles, builds a tiny dam across the neck. When the tide recedes, she breaks the dam to allow the air in the nesting chamber to be replenished.

MANGROVES

While salt marshes tend to be associated with temperate mud flats, mangroves dominate intertidal mud flats on tropical and subtropical coasts. There are many species, red mangroves (*Rhizophora mangle*) and black mangroves (*Avicennia germinans*) being the most common, and they provide shelter for young fish and other marine life, such as fiddler crabs (*Uca*) and the little fish (such as *Periophthalmus*) called mudskippers.

Wandering crabs have been particularly successful in exploiting many of the niches mangroves have to offer. Some have taken to the trees, thus avoiding the period of inactivity forced on fiddler crabs and other mud-dwelling organisms when they hide in burrows as the tide comes in. One climber is the mangrove crab (*Aratus pisonii*). It has sharp claws with which it can race along branches, even vertical ones. It gains water from dew and rain, and has little need to go down to the sea until spawning time.

Whether in the branches or on the mud below, food is normally plentiful amongst the mangroves, and the tangle of roots affords the crabs protection against predators. There is, however, one crab that has turned predator. The Caribbean mangrove crab (*Goniopsis*) lives in the mangroves of the Panamanian coast. It is coloured like a leopard and behaves like one; it stalks fiddler crabs and can pounce on them from about 1 ft (30 cm) away. If it cannot catch one it will settle for carrion, seeds, dead leaves and even mangrove mud. This opportunist roams far and wide, grabbing anything

GIANT CLAW *A male fiddler crab displays a single large pincer on the low-tide muds in a mangrove swamp.*

RED MANGROVE: A LONG-RANGE TRAVELLER

In estuaries and lagoons throughout the Caribbean, on the Pacific shores of Central America and down the coast of West Africa, the red mangrove (*Rhizophora mangle*) proliferates. It is a long-distance traveller, and its secret lies in its seeds. In spring the yellow flowers open, and after a few weeks these fall away to reveal a single seed underneath each flower.

The seed germinates while it is still on the tree; the seedling sprouts to a length of 6-12 in (15-30 cm) before dropping to the ground. If it drops like a dart into the mud, the seedling puts down roots immediately, but if it plunges into the sea it floats horizontally – that is, the whole length of the seedling floating on the surface of the water. As it drifts, however, the root tip slowly becomes waterlogged and sinks and the seedling gradually flips into the upright position. In this way it drifts for weeks or months in the ocean currents, even crossing the Atlantic Ocean. It can survive for about a year like this – the roots and top growth sometimes sprout while the seedling is still adrift. If it runs aground, it can take root and grow rapidly up to 2 ft (60 cm) tall in a single year. Prop roots raise the tree above the sea and trap silt.

Red mangrove is the pioneer species, and once a stand is established both black mangrove (*Avicennia germinans*) and buttonwood (*Conocarpus erectus*) move in. Soon all kinds of life is making its home among the mangroves. Fungi attack the insides of the mangroves' waxy leaves even before they fall. Once the leaves have fallen, bacteria and protozoans join them, forming a slimy brown film. Nematode worms, marine worms, flatworms and copepods feed on the film, while crabs and amphipods eat the leaf itself. Gradually it is broken down, its constituent parts eaten and re-eaten, the detritus providing food for many small organisms. These, in turn, fall prey to small fish, such as grunts, snappers and pinfish, which are eaten by larger fish. Slowly, an entire food chain is formed, making the mangrove a highly productive piece of natural marine real estate.

Tree Island ***Red mangroves form an island off Colombia's Caribbean coast.***

that looks edible, including the tails of feeding rats. *Goniopsis* itself may fall prey to a second mangrove predator, the slightly larger *Callinectes exasperatus*, another crab common in Caribbean mangroves.

Shark Nursery

Mangroves are nurseries for many species of marine animal not usually associated with shallow coastal waters. One such visitor is the lemon shark (*Negaprion brevirostris*). Unlike most other fish, female lemon sharks give birth to 'live' young – their 'pups' hatch out within the female. Lemon sharks give birth to their pups in the horseshoe-shaped Bimini atoll of the Bahamas, and the newborn youngsters head immediately for the safety of the mangroves. Here, during their first year, they patrol a small 1300 ft (400 m) by 130 ft (40 m) section of mangrove shoreline, living on a diet 80 per cent of which consists of small fish, such as young snappers and grunts, and 20 per cent of which consists of marine invertebrates, such as worms and shellfish. By the time they are about 32-39 in (80-100 cm) long, the young sharks will eat small octopuses, if they can catch them. As the sharks grow they move to a different area of the lagoon where they have increasingly larger territories. At 8 to 10 years old they move into more open reef habitats, and at 12 they move to the deeper waters of the Bahamas, working the reefs down to about 160 ft (50 m). Only the females will return to Bimini, to give birth.

Birthing Pool ***A female lemon shark prepares to give birth in a shallow lagoon surrounded by mangroves at Bimini off the Florida coast.***

Ice-cold Living

Ice may seem an unlikely place to live on or under, but for animals and plants inhabiting polar regions it is a key junction between land and sea. Penguins, seals, sea spiders, ice fish, krill and algae all manage to thrive in this environment.

Ice forms a permanent collar around the Antarctic continent. In summer its area amounts to a comparatively modest 1.5 million sq miles (4 million km^2). In winter, however, it expands to 7.7 million sq miles (20 million km^2) – the size of Europe. The first signs of the massive freeze-up to come are tiny crystals of pure ice, known as 'frazil', that form in the sea as ocean temperatures drop early in the Antarctic autumn. These soon become hexagonal discs nearly 1/8 in (3 mm) across and 1/32 in (1 mm) thick. Waves and swell push the growing crystals into pancake shapes, and these start to amalgamate into continuous sheets of ice. The sheets become huge ice fields, which despite their size are constantly shifting and cracking under the influence of changes in the weather, atmospheric pressure and oceanic motion. They fracture into floes (fragmented segments of sea ice), many of which are then driven together again to form hummocks and ridges.

Ice is, in effect, an extension of the land in the Arctic and Antarctic regions. Sea ice exists as fast ice (sea ice that is still attached to the land) and drifting pack ice. Freshwater ice is present as glaciers and icebergs; it also exists as vast ice shelves where continental ice sheets have been pushed out over the ocean and are floating on top of it – an example is the Ross Ice Shelf, the world's largest body of floating ice (about the size of France), lying at the head of Antarctica's Ross Sea. All these kinds of ice provide yet another living space for wildlife. Animals arrive from land, sea and air to take advantage of the opportunities offered in the zone where ice meets sea.

This includes the zone underneath the ice. More than 100 species of algae have been found in the holes, known as ice pockets, that pit the underside of Antarctic pack ice, rather like the holes in Swiss cheese. Comparatively little sunlight penetrates the translucent ice, but it is enough for the algae, which are able to function in 1 per cent of the light that penetrates open surface waters. As well as algae there are bacteria and algae-grazers such as the shrimp-like krill *Euphasia*. For the krill these ice pockets are a bountiful winter refuge. Denied free-floating phytoplankton in the open ocean, they gather amongst the ice and graze avidly. A single individual can clear the algae off an 8 in (20 cm) square portion of ice in just five minutes.

ANTARCTIC SURVIVOR *This small translucent crustacean is one of many creatures which thrive in the ice-cold Weddell Sea, Antarctica.*

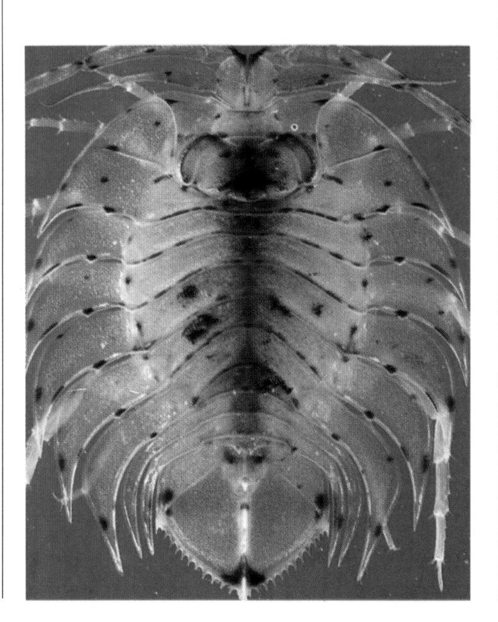

Antarctic Freezer

Life also exists under freshwater ice shelves. In some ways this is even more remarkable, for while sea ice is no more than 10 ft (3 m) thick in winter, ice shelves can be 230 ft (70 m) thick. Yet near White Island in the middle of the winter limits of the Ross Ice Shelf, a well-fed population of Weddell seals (*Leptonychotes weddelli*) lives about 12 miles (20 km) from the nearest open water. This suggests that the seals must be obtaining their food from under the ice. Researchers have cut boreholes in the ice and lowered television cameras through them. The results revealed an abundance of crustaceans and fish. Food reaches these unlikely habitats when water currents below the ice waft in particles from the open sea. These are eaten by the crustaceans, such as copepods, krill, mysids, amphipods and isopods, and they, in turn, are devoured by the fish.

Though growth is slow and in short annual bursts linked to the seasonal availability of food, there are a surprising number of 'giant' animals in Antarctic waters. There is, for example, *Glyponotus*, a giant isopod (a member of the order of crustaceans that also includes woodlice) which is one of the scavengers of the seabed. Being large has the advantage that offspring can be produced at a more advanced stage in the breeding season. The eggs of Antarctic species tend

ANTARCTIC ICE-SCAPE *Sea ice and icebergs fill the sea off the Antarctic Peninsula.*

to be larger and filled with more yolk than those of other species. This means that the embryos are better developed on hatching than those of the isopod's temperate or tropical relatives, and they have a better chance of survival.

Animals that grow slowly usually live longer than their faster-growing counterparts, so many of the animals living on, in or under the ice are long-lived. Antarctic sponges – one of which grows to over 3 ft (1 m) tall – are thought to be several hundreds of years old.

Normal limpets, barnacles and other marine creatures associated with rocky shores are not present on the exposed rocks of the Antarctic – moving ice makes it hard for anything to remain attached to the rocks. The exception to this rule is the Antarctic limpet (*Nacella concinna*) which emerges from the sea to graze on the mats of algae carpeting the rocks in summer. It covers itself with a special mucus that helps to prevent too many ice crystals from forming on it in the ice-cold air. It has the mats of algae to itself, for there are no other grazers here. Only kelp gulls (*Larus dominicanus*) swoop in to pluck the limpets from the rocks.

While the exposed rocks are sparsely inhabited, the sublittoral zone – the zone below the waterline – is rich with life, including various marine worms, sea slugs, sea cucumbers, sea urchins, sea spiders, starfishes and sea squirts. On rocky underwater outcrops there are bright yellow catkin-like soft corals, red, purple and orange sea anemones, some about 4 in (10 cm) across, and many species of pale brown or yellow Antarctic sponge – some of which mould themselves into the shape of the rock while others are free-standing. White brachiopods (lamp shells) – an ancient group of living fossils that dominated the sea about 550 million years ago – hang from bright red muscular stalks. There are even patches of red and green algae and lichens down here, grazed by red sea urchins.

CRABEATERS AND KRILL

Despite its name, the crabeater seal eats krill, not crabs. It has special interlocking teeth, which strain krill from each mouthful of water, in a similar way to the whalebone sieves of baleen whales. Like the great whales, it is a filter feeder. It feeds at night when the krill are close to the sea's surface.

SEX AND SPIDERS *Wisps of sperm and eggs are visible when Antarctic limpets (left) spawn on the sea floor. The giant sea spider (right) is another creature of the Antarctic.*

Sedentary animals, such as bryozoans, fall prey to sea spiders – not true spiders, but distant relatives. Sea spiders appear to have no bodies, although at the centre of the long legs is a short neck and a head with four eyes. They can have eight, ten or twelve legs, depending on the species. The body is so small that part of the gut and female reproductive tract extends into the legs. Moving very slowly, as if on tiptoe, a sea spider plucks bryozoans and sponges from the rocks with a large proboscis which is activated by its own ganglion of nerves, a second 'brain'.

All animals living in Antarctica must adapt to the cold conditions, and in the inshore waters they have found several ways in which to survive, one being the method of the Antarctic ice fish (family Channichthyidae) which include *Pseudochaenichthys georginus*, with a head that takes up a third of its 28 in (71 cm) body, a snout like a duck's bill and a large mouth edged with sharp-toothed bony lips. Ice fish lower the freezing point of their blood and other body fluids by manufacturing a chemical antifreeze. This prevents their tissues from being shattered by ice crystals, in the same way that antifreeze protects the water radiator of a car in winter.

For ordinary fish, another problem in such icy waters is their haemoglobin – the red iron-based molecule that carries oxygen round the body of most vertebrates and some worms. It is inefficient at low temperatures. The ice fish get round this difficulty by having no haemoglobin, save for a few vestigial corpuscles. The result is a ghostly, ivory-white fish with thin, almost colourless blood, an outsize heart and wide blood vessels which produce a blood circulatory system capable of moving large volumes of blood at low pressure. The price the ice fish pays for this adaptation is that it is slow and barely able to compete with its red-blooded relatives. Therefore it is confined to the very bottom of the ocean – a place where the ice fish is king.

EMPEROR OF THE SOUTH

Antarctica's surface-dwellers tend to be seasonal, visiting the frozen continent only in summer. There are, however, some animals that survive and thrive on the ice even in winter. They do so not with special body fluids but with a thick layer of subcutaneous blubber – up to a third of the body weight in the case of penguins – and a dense covering of feathers or fur. These allow them to endure the most extraordinary conditions.

No story is more remarkable than that of the emperor penguin (*Aptenodytes forsteri*), which is exposed to the most extreme weather conditions of any known vertebrate. Every year the adult birds, which stand 45 in (115 cm) tall, make their way across more than 340 miles (550 km) of pack ice until they reach their breeding sites on solid fast ice, close to the Antarctic mainland. They either walk the journey or toboggan it on their stomachs, and they do it not in spring – like all other migrating Antarctic and subantarctic animals – but at the end of the southern summer, in April and May. So, while the rest of the community

NON-FROZEN FISH *An Antarctic ice fish appears pale pink because of its low levels of red blood cells. It can survive in an icy sea by having antifreeze in its blood.*

SAFETY IN NUMBERS *Male emperor penguins, each with an egg balanced on its feet, group together for protection against the swirling midwinter blizzard.*

heads north, away from the South Pole, the emperors travel south towards it.

When they reach the breeding site, males and females enter an intensive five-week period of courtship. Unlike other penguins, emperors usually have different partners each year, although they will stay loyal to their chosen mate for the entire breeding season. A key ritual during courtship and mating involves the pair stretching out their necks, bowing and braying loudly at each other. Between May and early June, just as the winter weather begins to close in, the female lays a single 5 in (12.5 cm) long egg, which she passes immediately to the male. He balances the egg on his feet, and covers it with a thick fold of warm abdominal skin. The female leaves him to it, setting off for the coast, where she will feed during the worst of the early winter weather. The male stays behind to incubate the pair's precious egg.

During blizzards, all the male emperors at the breeding site huddle together for warmth, each penguin resting his bill on the bird in front. There may be as many as 5000 birds present, the birds at the centre of the huddle using less energy to stay warm than those on the edge. It has been calculated that birds in the middle of the group use up 3½ oz (100 g) of stored fat per day, whereas those in the open use up twice as much. Birds on the outside, therefore, try to shuffle to the warmest places at the centre of what turns into a constantly revolving circular parade. During a 48-hour blizzard one huddle was seen to move 660 ft (200 m). Astonishingly, no fights break out, which suggests that the emperor must have found a way to suppress otherwise aggressive penguin behaviour. When a storm passes the group splits up, and each penguin maintains not quite a territory, but his own personal space.

LIVING TOBOGGAN *Adult emperor penguins journey vast distances across the Antarctic ice, travelling to and from their inland breeding grounds and the sea.*

SPRING CHICKS *An emperor penguin chick sits on its father's feet (left) anticipating its mother's return from a winter fishing at sea. A crèche of half-grown emperor chicks (right) huddle together while waiting for their parents to return with food.*

Male emperors spend more than 60 days of the harshest weather experienced on Earth incubating their eggs, the only polar animal to perform such a feat. Temperatures plummet to below –40°C (–40°F), winds reach speeds of over 125 mph (200 km/h), and ice crystals are blown like stinging pellets, yet the egg protected inside the male's brood pouch is kept safe and sound.

The 6 in (15 cm) tall chick hatches out in mid-July, and the female times her return to coincide with its emergence. Should she be delayed, the male is able to feed his chick with a special 'milk' – a gullet secretion rich in proteins and fats – and he is able to do this despite his 60 day fast. The female's arrival means a supply of freshly regurgitated fish and squid for the chick, and she takes over brooding duties. The male, near-starving and weak from his long fast, trudges slowly towards the nearest polynya – a large patch of open water in the pack ice. The journey may take up to two weeks across 80 miles (130 km) of ice.

The birds are superb aquanauts. With a streamlined body designed primarily for a life in the sea, the emperor can descend at about 6½ ft (2 m) per second, hunting for fish and squid down to depths of 1525 ft (465 m), making it the deepest-diving penguin. It can also stay down for as long as 15 minutes. Yet, despite these abilities, emperors are not efficient hunters. In normal circumstances these 29 lb (13 kg) birds require 6 lb (2.5 kg) of fish or squid each day just to keep going; when looking after its offspring, an emperor must catch an additional 7 lb (3 kg) each day to feed the youngster. On each fishing trip, then, an adult must catch between 50 and 90 fish or squid, each weighing 5 to 7 oz (150 to 200 g). During the five to eight days it spends chasing its food, an individual may make 500-1200 dives, of which 10 per cent will be successful: it will catch something during only 50-120 dives.

By the beginning of the southern spring, all the six-week-old chicks have gathered in crèches. The parents no longer have to travel so far, because the sea ice has started to recede, and they establish a shuttle service between the sea and the breeding colony. Even so, round trips of up to 1028 miles (1654 km) have been recorded. The returning parent finds its chick by listening for its distinctive voice in the crowded crèche.

By December the chick has discarded its thick, brown, downy coat and completed its moult into adult plumage. It fledges with only about 60 per cent of the adult weight, but the parents stop feeding it. Eventually hunger forces the young bird to go to sea in order to obtain food.

MAINLAND PENGUINS *Adélie penguins examine a potential nest site.*

ADÉLIE PENGUINS

Adélie penguins (*Pygoscelis adeliae*) travel even farther to the south than the emperors. They go as far as the Antarctic mainland, making them the penguins that breed closest to the South Pole. Unlike the emperors, however, they do it in the southern spring and summer, not in winter. The males arrive first, in September and October, each one finding a nest site and declaring his occupation with a display in which he stretches his neck, claps his bill and flaps his wings. Any intruders are quickly seen off. When the females arrive, a few days later, a similar performance attracts a partner, usually the same one as in the previous year.

Taking the Plunge *Adélie penguins back up on the ice until one penguin makes a move and they all jump in.*

The pair then settle down to nest-building.

The nest is simple. It consists of a platform of stones with a shallow hollow in the top, which ensures that its occupants are high and dry above any melting snow or ice. The mother lays two eggs in October or November; both parents share incubation duties. The chicks hatch by late December.

Skuas swoop in to seize unguarded chicks and feast on unattended eggs, so the Adélie parents have to be constantly ready to fend them off as best they can. During the chicks' third week after hatching, they join a crèche, each containing about 20-30 birds, which offers them some protection both from predators and against the cold. The parents, meanwhile, trek to and from the sea, returning every couple of days with their crops full of krill, fish and squid.

Eventually, in mid-February and March, the young birds complete the moult from chick-down to adult plumage, and they head for the sea. At the edge they all stop in their tracks, no bird wishing to be the first to jump. The pile-up of birds means, however, that sooner or later an individual is nudged and falls into the water. The others follow. Waiting for them are opportunistic leopard seals (*Hydrurga leptonyx*), which seem to sense when the time is right to patrol the penguin slipways. They appear from nowhere, the first sign being a large pair of powerful jaws wrapped around an unfortunate victim. The seal thrashes the bird about on the water surface in order to strip the meat off the bones, and then swallows its meal whole.

Antarctic Seals

There are four species of seal that live permanently in, on or under the southern polar ice: the crabeater seal (*Lobodon carcinophagus*), which feasts exclusively on krill and is the most common seal on Earth, with a total population of 15-30 million; the leopard seal, which feeds mainly on its cousins the crabeater seals as well as penguins, fish, krill and other prey; the Ross seal (*Ommatophoca rossi*), which lives on the deep pack ice and is believed to feed mainly on squid; and the Weddell seal (*Leptonychotes weddelli*), which lives in the denser parts of the pack ice, feeding mainly on fish.

In spring, as the sea ice breaks up and the breeding season comes round, each species has its own preferences for a habitat. The female crabeater seals haul out on small ice floes, one female to each floe. A little later a male seal hauls out on the same floe. The female is pregnant from mating the previous year, and gives birth to a single pup which she keeps close by her. Even though the male is not the pup's father he inadvertently protects it and its mother when defending what is now his territory – the occupied ice floe. He sees off other male crabeaters and even dangerous leopard seals.

The female is indifferent to the male at first, for she is interested only in raising her pup, but as it grows the male seems to become impatient. He tries to mate with her,

Underwater Predator *Leopard seals patrol the sea close to penguin colonies and grab victims as they jump off the ice and into the sea.*

COMMUNITY LIVING *Outside the breeding season, crabeater seals congregate on the same ice floe.*

but is rebuffed, often violently. Eventually he manages to get between the mother and pup, and this would be the logical time for mating to take place, though no one has yet witnessed the event.

The young crabeaters – which grow rapidly and are weaned in less than four weeks – enter the water when only a month old, having finally been abandoned on the ice by their mothers. It is a dangerous moment. Leopard seals and killer whales (*Orcinus orca*) are sometimes waiting for them – about 80 per cent of adult crabeaters bear the parallel scars of leopard-seal attacks on their bodies. However, if a young one survives its first year unharmed, it can expect to live for another 40 years. After that first year, crabeaters are usually able to spot and outmanoeuvre leopard seals, though killer whales remain a more serious threat. They take seals of all species and ages, organising themselves into well-ordered hunting packs, and even tipping them off ice floes by buffeting the ice from below. Most of the animals they target do not survive.

UNDER THE ICE *Weddell seals raise their pups in cracks in the ice. Groups discourage interlopers from entering their territories.*

Unlike their crabeater cousins, female Weddell seals form into small pupping colonies. When the time for breeding arrives, between September and November, they haul out of the sea onto large, stable ice platforms, emerging through tide-cracks that always occur in the same places, close to the continent. Each female gives birth to a single pup, and suckles it for five to six weeks.

The females form groups dominated by a single male. The male establishes a three-dimensional territory that includes the water below the ice platform as well as the platform itself, and this he defends against all comers, using threatening calls. If intruders enter the territory they receive short shrift. Fights can be bloody, and the resident male is usually covered with many wounds and scars. Just before weaning their pups the females come into heat, and mating takes place in the water below the platform.

Weddell seals rarely live beyond 20 years old, partly because of tooth wear. The only way they can survive on the pack ice in winter is by keeping breathing holes open using their teeth. Consequently, their teeth wear down and are open to infection. In old age this condition prevents them from keeping the holes open, and they die.

MOMENTARY MOTHERHOOD

At the other end of the globe, Arctic harp seals (*Phoca groenlandicus*) and hooded seals (*Cystophora cristata*) live in large groups on the pack ice near Labrador in Canada, near Jan Mayen Island off Norway and in Russia's White Sea. Both species breed on the ice, but they have even shorter weaning periods than their Antarctic cousins.

Indeed, the hooded seal has the shortest weaning period of any known mammal: from birth to weaning in just four days. It sheds its lanugal (prenatal) hair while it is in the womb, so it is born with a short-haired, blue-grey coat. Its mother produces very rich milk, with high levels of butterfat,

ON THE LOOKOUT *A harp seal surfaces through a breathing hole in Arctic mush ice and looks for signs of danger.*

Red Balloon *A male hooded seal sees off rivals with an aggressive display: it blows out its red nasal septum into the shape of a balloon.*

so that during four days the pup is able to double its body weight. Hooded seals form small family groups for the breeding season, each group establishing itself on an ice floe. After weaning, each female comes into heat, and the male, who has been waiting nearby for the rearing process to be completed, mates with her.

During the rest of the year hooded seals remain offshore, feeding on Greenland halibut (*Reinhardtius hippoglossoides*), redfish (*Sebastes marinus*), capelin (*Mallotus villosus*) and Arctic cod (*Boreogadus saida*). Harp seals feed mainly on fish, capelin being their favourite food in the south of the range and Arctic cod in the north. They also eat shrimp-like creatures similar to the Antarctic krill. Ringed seals (*Phoca hispida*) – the most numerous seals in the Arctic – also feed on small krill-like crustaceans.

The largest of the Arctic seals is the walrus (*Odobenus rosmarus*). It lives in two distinct populations – one in eastern Siberia and Alaska, and the other in eastern Canada, Greenland and northern Europe. It forages in shallow water, sometimes under the ice, feeding on Arctic clams (*Mya truncata*), which it detects and extracts from the mud with its heavy lips and stiff whiskers. It used to be thought that this was the walrus's staple food, but recent studies have shown that it is partial to other animals, too. Researchers from the University of Alaska examined the stomach of a walrus and found a chunk of ringed seal meat, a piece of a black guillemot and the siphon of a soft-shelled clam, suggesting that the walrus may have a surprisingly catholic diet.

Walrus pups are born on the pack ice in May, and despite the great size of their parents they are vulnerable to predators. In fact, all species of Arctic seal fall prey to hunters. There are killer whales, like those in the south, and there are opportunists, such as the Arctic fox (*Alopex lagopus*), which can tackle a young pup. But there is one supreme hunter which lives in the Arctic and tackles just about anything – the polar bear (*Ursus maritimus*).

Arctic Wanderers

The magnificent polar bear rivals the Kodiak bear (*Ursus arctos middendorffi*) – a kind of grizzly found on Kodiak Island off Alaska – as the world's largest living carnivore. It is

Old Whiskers *A male walrus (above) arrives at a hauling-out beach in Alaska. Right: Walruses rest on an ice floe at the edge of pack ice near Baffin Island, Canada.*

ICEBERGS – FLOATING ISLANDS

Each year about 15 000 Arctic icebergs calve – break away from their mother glaciers, most of which are in Greenland. They then drift south in the slow-moving Labrador and East Greenland currents. The above-water portions of some are the size of cathedrals, up to 560 ft (171 m) high, carved by the sea into fantastic, subtly coloured shapes – deep blue caverns, grey arches and ice cliffs tinged green by algae.

Their surfaces glow soft pink, orange and purple, taking the colour of the sun and squeezing it between the tiny tightly packed crystals of ice.

Down their sides ribbons of meltwater trickle into the sea. The heavier fresh water empties into the lighter salt water, driving an upwelling of nutrients around the margins of the bergs. 'Ice-shrimps' – large free-swimming crustaceans – crowd the less saline water around the base and are food for small fish that dart in and out of an iceberg's shadow.

At the other end of the world, flat-topped tabular icebergs break away from the ice shelves of Antarctica during the southern summer. Many are small cottage-sized bergs called 'growlers'. However, some Antarctic icebergs are truly enormous. In 1967 an iceberg from the Amery Shelf crashed into the Fimbul Ice Shelf and broke off another iceberg with an area of 3030 sq miles (8000 km^2). This became known as Trolltunga, and it survived for 12 years until 1979, when it fragmented and disappeared off South Africa.

About 80 per cent of an iceberg lies hidden below the sea's surface. When the sea has eroded the base and the berg becomes top-heavy, it turns turtle, exposing the spectacularly iridescent greens and blues of the many species of algae which have been growing on the underside.

SCULPTED BY THE SEA *An iceberg, showing the fluted grooves caused by wave erosion, is left high and dry by the tide.*

a nomad that wanders the Arctic in search of prey, mainly seals, creeping up on them as they rest on ice floes. Submerging without making a sound, it enters the water some distance away from the floe and then re-emerges beside it, smashing the seal's skull with one blow from its forepaws. The bears also wait beside seals' breathing holes, ready to snatch a prey.

The polar bear is a true Arctic animal, protected from the cold air and water by a thick 4 in (10 cm) layer of blubber below the skin and a warm, water-repellent, two-layered fur coat covering the entire body except the nose and feet pads. The dense woolly underfur is such a good insulator that very little heat escapes from the body. The erect hairs of the outer coat are hollow, and, like row upon row of tiny double-glazed windows, they trap the heat from the sun,

ON THE HUNT *A polar bear follows the scent of a seal below the Arctic ice.*

Natural-born Swimmers *Polar bears are truly marine mammals. Tired water-borne cubs like these may place their front paws on mother's back in order to rest.*

warming the polar bear's black skin. Any heat radiating from the skin is reabsorbed by the hollow hairs. In addition, the hair focuses ultraviolet light reflected off ice and snow onto the skin, and the energy absorbed is turned into more heat. The conversion is 95 per cent efficient. By soaking up and recycling heat in these ways, the polar bear sometimes has a skin temperature that is warmer than the inside of its body, even when the air temperature is well below 0°C (32°F). The bear is, in effect, a living solar collector.

Male bears are solitary, while females usually have a couple of cubs in tow. Unlike most carnivores they cannot have a permanent home range, for the ice on which they live is in constant motion, and the position of seal concentrations is difficult to predict. The result is that bears tend to congregate where seals are plentiful, and then the males fight for dominance. The battles can be quite violent – the fresh wounds and broken teeth found on many male bears are witness to the ferocity of their conflicts. A large, dominant bear – twice the size of the female – then latches onto a female on heat and stays with her for a few days.

All this happens in spring, the bears' mating season. After that, they part company. The female spends the summer hunting among the pack ice, and then retires to a snow den for the winter. In January the cubs, usually twins, are born in the den, and they feed on their mother's milk until they are ready to emerge in March or April.

Not knowing where the next meal is coming from in the vastness of the Arctic ice, polar bears will take advantage of any animal, even turning cannibalistic. On Southampton Island at the mouth of the Hudson Bay, observers watched a fight between two adults – one male, the other female. The male tried to take the female's cub from her. She fought back in defence of her cub, but the male was the victor and proceeded to feast on the female's body.

THE PRODUCTIVE ARCTIC

While the coastal waters of the High Arctic are relatively barren, parts of the Low Arctic are rich in nutrients and therefore rich in marine organisms. It has been estimated that in summer the Barents Sea, the shallow section of the Arctic Ocean north of Norway and western Russia, contains 98 million tons of zooplankton, mainly the crustaceans *Calanus*. These feed on an annual total of about 2-3 million tons of plant plankton.

Food in the Icy Wilderness *A solitary polar bear on the Arctic ice sniffs the air for signs of a meal.*

4 Surrounded by Sea

Island Aristocrat *A king penguin performs its 'ecstatic' mating display.*

Every island – and there are more than half a million of them scattered across the Earth's oceans – is an ecological world of its own. From the volcanic Galápagos Islands of the eastern Pacific, to the coral atolls of the South Pacific, to the bleak and windswept isles of the subantarctic, plants and animals have developed and evolved in ways that are unique. The Galápagos have their tool-wielding finches. Shearwaters and noddies colonise coral cays, while colourful arrays of fish swim through the surrounding reefs. In the breeding season, penguins and seals congregate in their thousands or even millions on the subantarctic islands.

Predator *The white-tipped reef shark hunts by night.*

VOLCANIC ISLANDS

Born of great heat from the molten upper mantle of the Earth's core, volcanoes punch through weaknesses in the crust on the sea floor and push through the ocean surface as isolated islands. Life is quick to find them.

Once they were barren rocks that volcanic activity had pushed up above the ocean surface. Then, gradually, they began to acquire life. Seeds, fruits, spores and the eggs of invertebrates arrived, blown by the wind or floating on the sea. Migrant birds were blown off course and made landfall on the new volcanic islands. Floating debris brought large spiders, centipedes, reptiles and even land mammals that had somehow got cast adrift from the continents. Before too long, populations of plants and animals had become firmly established.

They also began to strike out in new evolutionary directions. Among the pioneering inhabitants of the volcanic islands, the plants and animals that adapted well to particular habitats survived; ones that were less adaptable did not. Species eventually emerged that were quite different from the original colonisers. About 25 per cent of the species found in the coastal waters off the Galápagos Islands in the eastern Pacific, for example, are endemic – that is, they are unique to the islands and the zone around them. Surprises the Galápagos have to offer include cormorants that have lost the ability to fly and penguins – normally associated with much colder Antarctic or subantarctic climes – living at the Equator.

TOOL-USER *The woodpecker finch uses a cactus spine to winkle out grubs from below tree bark.*

GALÁPAGOS MARVELS

The unique wildlife of the Galápagos Islands, lying about 680 miles (1100 km) west of Ecuador, occupies a special place in the history of science, for it was Charles Darwin's observations of it that helped him to develop his theory of evolution. Visiting the islands in 1835 he noticed, among other things, the numerous species of finch living there. Each is highly specialised, including finches that drink blood, some that eat insects and others that use a tool to help them to feed – a cactus spine for extracting grubs from tree bark. These little birds, descended from a basic seed-eating finch-like bird that probably flew in from South America, have evolved into all sorts of forms, each of them adapted to a different way of life. Isolation on the new islands, and the different demands of each environment in which the birds found themselves, gave rise to great diversification.

Life along the Galápagos shoreline has been even more prolific in specialist adaptations. One unique creature is the marine iguana (*Amblyrhynchus cristatus*), a reptile that, in the absence of mammalian predators, has been able to dominate the rocky volcanic coasts. It has evolved a blunt snout for grazing on seaweed, a laterally flattened tail to help it to swim and powerful limbs with strong claws to help it to cling to rocks and resist being dislodged by the waves. Despite its name, the Galápagos marine iguana spends 95 per cent of its time on land. This is because the Galápagos Islands, although straddling the Equator, are washed by the cold waters of the Humboldt Current. When feeding in them the marine iguana loses body heat rapidly – up to 17°C (30°F) during a dive. It has to bask in the sun to raise its body temperature to 36°C (97°F) before returning to the sea to feed. While it basks, poised as still as a statue on its roosting rock, Galápagos mockingbirds (*Nesomimus parvulus*) fly down to pick off parasites and fragments of dead sloughed skin.

If the cold waters are a problem for the iguana, they are ideal for other marine creatures, including the Galápagos sea lion

SHALLOW FEEDER *The Galápagos marine iguana swims in shallow water in order to reach the green algae on which it feeds.*

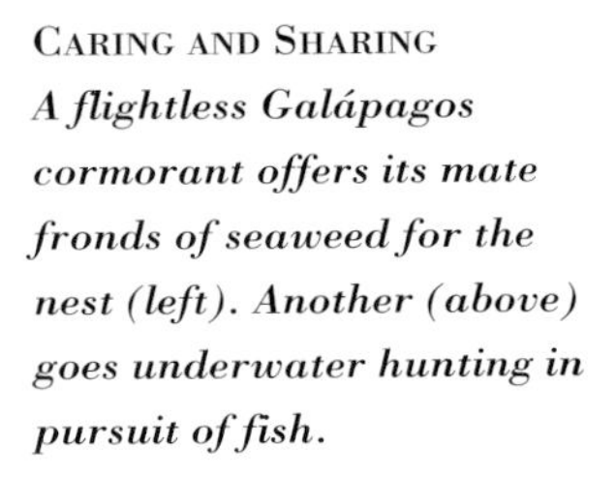

Caring and Sharing *A flightless Galápagos cormorant offers its mate fronds of seaweed for the nest (left). Another (above) goes underwater hunting in pursuit of fish.*

(*Zalophus californianus wollebaeki*), Galápagos fur seal (*Arctocephalus galapagoensis*), Galápagos cormorant (*Nannopterum harrisi*) and Galápagos penguin (*Spheniscus menidculus*). All these animals belong to families that are more usually associated with colder climates in higher latitudes.

Among them, the Galápagos cormorant has taken its family's ability to pursue fish under water to such an extent that it has lost the ability to fly. It uses what is left of its wings not to flap, like a penguin or an auk, but to steer. Nesting pairs have another curious habit. Every time the parent birds change incubation or brooding shifts, they greet each other in a most courteous manner. The returning bird calls from the water and is answered by its mate on the nest. As it leaves the water, it picks up a tuft of seaweed or a yellow starfish and, bowing deeply, it presents the gift as if offering a bouquet of flowers or a box of chocolates. The partner receives the gift and lays it gently at the edge of the nest. After drying its wings in the traditional way, the returning cormorant then picks up a small stick and again presents it to the bird on the nest. It is duly received and placed at the nest edge, and then the two birds change places. The other bird makes for the sea, while its newly arrived mate takes over brooding duties.

How to Stop Overheating

In the cool, fish-rich water the Galápagos's seagoing birds and mammals are in their element, but when they emerge onto land, unlike the iguanas, they face the problem of overheating. For the Galápagos fur seal this has led to small size: it is the smallest marine mammal in the world. Compact size is an adaptation to reduce overheating – the smaller the surface area, the less an animal overheats. It is an example of Bergmann's Law, according to which closely related animals tend to be smaller towards the tropics: so the emperor penguin, which lives close to the South Pole, is the largest penguin, whereas the Galápagos penguin, living on the Equator, is the second smallest. The Galápagos sea lions show other adaptations. They stay in the water during the heat of the day; they even mate in the water rather than on land.

The islands are a popular breeding site for birds that may spend much of their life away from the Galápagos. In the low scrub behind the rocky shore, two species of frigate bird, the magnificent frigate bird (*Fregata magnificens*) and the great frigate bird (*F. minor*), court and nest. The male inflates a huge, bright red, balloon-like throat pouch to attract the female. As the pair meet, they exchange

Tropical Seal *A Galápagos fur seal sleeps on the surface of the sea in order to keep cool.*

calls that sound like eerie laughter from a haunted house. The male frigate bird then produces a rattling sound from his throat and wags his bill from side to side.

The blue-footed booby (*Sula nebouxii*), a tropical relative of the gannets, also breeds on the Galápagos Islands. The male performs a stiff-legged dance during which he ensures that the female cannot fail to admire his large blue feet. On Española the chicks of a close relative, the masked booby (*S. dactylatra*), are sometimes injured by delinquent males that have not acquired partners; they try to mate with the young ones. The injured chicks are then attacked by one of the island's species of mockingbird, *Mimus macdonaldi.* This species has turned to drinking blood, harassing any bird with an open wound. The mockingbird keeps the blood flowing while it drinks its fill.

OUT TO IMPRESS ***The dramatic throat display by a male great frigate bird is meant to impress a potential partner.***

On Wolf Island the sharp-beaked ground finch (*Geospiza nebulosa*) torments adult masked boobies in a similar way. It pecks at the base of tail feathers, breaking quills and drawing blood. It is thought that the technique evolved at a time when the bird fed on the bloodsucking flies and lice that parasitise boobies. While eating the blood-filled parasites, the bird probably developed a liking for the blood itself. As if that were not enough, the ground finch also breaks into masked booby eggs and extracts the contents – a habit it shares with the Española mockingbird. It is too weak to pierce the eggshell, so it presses its beak against a rock to give it some leverage, and kicks the shell against a rock opposite until it cracks.

BODY TALK

Also living off the coasts of the Galápagos Islands, as off most of the world's shores, are some of the most sophisticated shoreline communicators. Imagine being able to soothe a mate, frighten off a predator, warn a rival and show that you are frightened – all at the same time. Cephalopods – octopuses, squids and cuttlefish – can do all that, with body language.

TREETOP SHORE BIRD ***A brown pelican parent guards its chicks in a nest on top of mangroves on Santa Cruz Island in the Galápagos archipelago.***

By changing the designs, patterns and colours on their bodies they can hold many conversations at once. In the skin of one of these animals there are thousands of elastic-walled colour cells that can be filled or drained of pigment within seconds. In fact these remarkable creatures are probably capable of the fastest changes of colour in the entire animal kingdom, involving dramatic transformations – longitudinal or diagonal zebra stripes, dark transverse bars, spots and blotches and moving patterns of colour that start at one end of the body and travel to the other. Biologists believe that these patterns give an insight into the workings of the animals' brains – a window on the cephalopod mind.

The story does not end there, for visual signals from a group of very simple but effective communicators – tiny red algae called dinoflagellates – also tune in to the cephalopod brain. If they are attacked by

HIDING PLACE *A Pacific manta ray has a shoal of smaller fish hiding under its giant 'wings'.*

predators, the dinoflagellates emit a luminescent flash. The squid or cuttlefish are not the predators, because the dinoflagellates are too small to be of any gastronomic interest to these far larger creatures, but the flash is still given to attract their attention. The dinoflagellates are eaten by small fish and shrimps, and so by attracting the cephalopods, which eat small fish and shrimps, they take the heat off themselves.

For other sea-dwellers, meanwhile, the Galápagos Islands are a beacon, meeting place and refuge in the Pacific Ocean. Offshore are pods of sperm whales (*Physeter catodon*) and huge, meandering schools of scalloped hammerhead sharks (*Sphyrna lewini*), surrounded by blue-and-yellow surgeon fish (*Xesurus laticlavius*). Inshore, great shoals of 70 or more golden rays (*Rhinoptera steindachneri*), each 2-3 ft (60-90 cm) across, brush the undersurface of the sea as they feed on plankton in the shade of a cliff. At night, Pacific manta rays (*Manta hamiltoni*) – the largest of all manta rays, up to 20 ft (6 m) across and weighing over a ton – turn somersaults as they, too, feed on the plankton.

LOW DIVERSITY

The Galápagos Islands have a wealth of unique species; on the other hand, the overall diversity of plants and animals is surprisingly low compared with the mainland. There are, for example, just 500 native plant species on the islands, as opposed to more than 10 000 in Ecuador. The islands are young, geologically speaking, having risen over a volcanically active 'hotspot' in the Earth's crust between 3 and 5 million years ago. The normally arid climate is slow to break down the volcanic rocks and slow to be colonised by plants. Fernandina, the youngest of the islands, has only about 10 per cent of its surface cloaked with vegetation. There are few habitats to colonise, so the coastline is generally bleak and rocky.

Also, the newer the islands, the less chance there has been for wayward plants and animals to chance upon them. It has been estimated that the present flora of the Galápagos Islands is the result of only about 400 chance arrivals, which means that, on average, one species has been established every 12 000 years or so.

Moreover, there is an evolutionary conundrum in the Galápagos. While the land-based finches and many of the plants could have evolved into their present forms in a

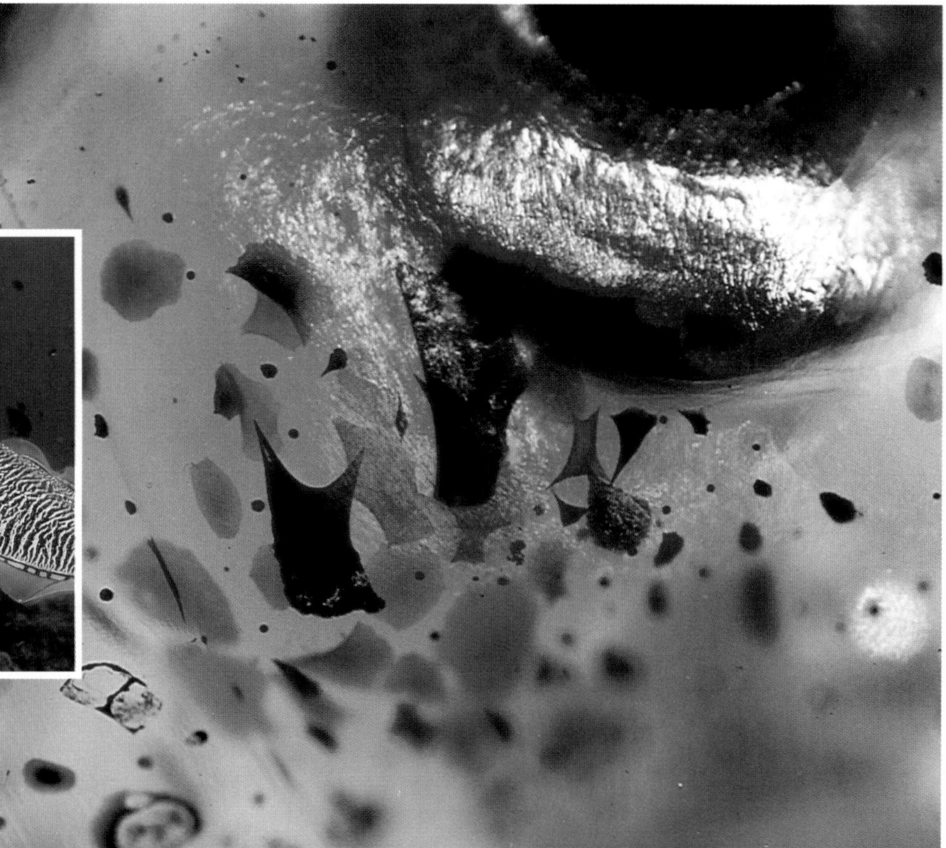

COLOUR CODED *Colour pigment cells (right) allow cuttlefish to send signals to each other (above) by changing the patterns on their skins, an ability they share with squid and octopuses.*

Isles of Fire *The islands of Hawaii owe their position, shape and volcanic activity to forces which erupt from the bottom of the sea.*

relatively short period of geological time (it has been determined that the finches could have evolved in less than a million years), the same cannot be said for the marine iguana and its land-based relative the Galápagos land iguana (*Conolophus subcristatus*). To evolve into their present forms, they would have had to arrive in the Galápagos about 15 million years ago. The islands, however, are less than 5 million years old.

Hawaiian Duck *The Laysan teal lives on the Hawaiian island of Laysan, where it eats flies.*

The answer to this puzzle was discovered in 1992, when a sea-floor survey by researchers from Oregon State University chanced upon seamounts (undersea mountains) with signs of wave erosion, such as rounded pebbles and wave-cut terraces. These suggested that the seamounts were once volcanic islands that had formed over what is now known as the Galápagos hotspot. They had been gradually shifted east, as if on a massive conveyor belt, by the movements of the Earth's crust. They were then drowned by the sea and replaced over the hotspot by the present Galápagos Islands. A common form of iguana might somehow have rafted from mainland South America to these earlier islands, and from thence to their successors.

The Hawaiian Islands

The Hawaiian island chain is the most isolated major island group in the world. Like the Galápagos it has formed over a volcanic hotspot. Also like the Galápagos, its isolation has resulted in the evolution of a unique flora and fauna.

On land, Hawaiian honeycreepers (family Drepanididae) have evolved from carduelinc finches (family Fringillidae, subfamily Carduelinae) in much the same way that the Galápagos finches evolved from their pioneering ancestors. Of the rest of the living things in Hawaii, about 90 per cent are endemic. These include a species of duck on Laysan Island – the Laysan duck (Laysan teal, *Anas laysanensis*) – that feeds on the swarms of flies that gather over saline ponds. On the islands' shores there are Hawaiian monk seals (*Monachus schauinslandi*), a primitive species thought to be similar to the ancestor of all seals. In the sea, schools of pregnant grey reef sharks (*Carcharinus amblyrhynchos*) patrol the nearshore reefs by day, and white-tipped reef sharks (*Triaenodon obesus*) feed amongst the coral heads by night.

Being so remote, Hawaii is a useful haven for animals that hunt in the ocean but need a safe place to rest. In small coves and bays, schools of spinner dolphins (*Stenella longirostris*) rest by day, 'sleeping' with first one side of the brain and then the other, while swimming effortlessly back and forth across the inlet. Come the late afternoon, however, they prepare to go to sea to feed. They leap from the sea, spin in the air and plunge back in with a loud splash. The spinning becomes more frantic as the evening draws in. It is as if the school is undergoing some kind of roll call, checking that all the dolphins are present and ready before they go to feed.

At some unknown signal the leaping ends, and the school heads for the open sea. Here it meets up with other 'day schools', and they merge into a large night-time hunting school, sometimes over 1000

Ancient Seal *The primitive Hawaiian monk seal is one of the rarest mammals on Earth.*

Dolphins by Day *Small groups of spinner dolphins 'rest' in sheltered bays amongst the Hawaiian islands by day. At night they go out to sea to fish in huge schools.*

strong. In the morning the dolphin schools return to their daytime resting bays throughout the Hawaiian Islands.

Another remarkable Hawaiian wildlife spectacle occurs each summer along the shore line of Laysan Island. Here the Laysan albatross (*Diomedea immutabilis*) and black-footed albatross (*D. nigripes*) have breeding colonies, each pair raising a single chick. In the colony the young birds are relatively safe, but when they leave the nest danger threatens.

The fledgling albatrosses unfurl wings that have a span of 5 ft (1.5 m). Then, like hang-gliders, they use the wind to take off and hover a few feet in the air, before dropping back down again. Gradually they gain more confidence, and when the wind blows more strongly, they fly farther and farther until they land not on the beach but in the sea. Then they must learn to take off from the water. This is an absolutely vital lesson, for waiting below the surface to seize them if they do not take off quickly are tiger sharks (*Galeocerdo cuvieri*).

How these sharks know the right time to arrive is a mystery, but each summer many of them come into shallow water and patrol the shoreline just as the largest number of fledgling albatrosses are leaving their nests. The sharks do not usually swim in such shallow water, where they have less room to manoeuvre than in the open; nevertheless, they take advantage of the sudden abundance of food at the surface. The birds, however, have an even chance of surviving a shark attack. A tiger shark usually dives down, and then attacks a victim from directly below, but in the shallows it is able only to sideswipe or nudge the prey at the surface. So the shark is forced to bring its head and back right out of the water, and the bow-wave it creates often pushes the bobbing albatross chick ahead of it, giving the bird an opportunity to get away.

In the western part of the Indian Ocean lie the Amirantes Islands, and on the small islet of Boudeuse there is one of the world's largest breeding colonies of masked, or blue-faced, boobies (*Sula dactylatra*). The birds feed by plunge diving, but not all return to the surface intact, and some do not return at all. Some have torn wings and missing legs. The predators responsible here are large fish – groupers (*Epinephalus tukula*) – which lie in wait for the birds and intercept them when they dive for smaller fish.

Christmas Crabs

On the eastern side of the Indian Ocean lies Christmas Island, covered with rain forests whose main animal inhabitants are land crabs. There are more than 20 species of these crabs, including the blue land crab

Tropical Albatross *This Laysan albatross chick (below) will grow into a magnificent adult (bottom) with a 6½ ft (2 m) wing span.*

BIRD-EATING SHARKS *At albatross fledgling time, tiger sharks gather inshore at Laysan to seize fledglings which have come down in the sea.*

(*Cardisoma hirtipes*) and the robber crab (*Birgus latro*). Most numerous of all, however, is the red crab (*Gecarcoidea natalis*), of which there are thought to be approximately 120 million individuals. They all live on the rain-forest floor, which they keep clear of leaf litter and other organic debris, including seedlings. How new plants germinate is a mystery, but the mature trees receive adequate compensation in the form of crab droppings. The crabs' burrows also aerate the roots.

The crabs spend most of their lives on land, but every year they have to go down to the sea to spawn, and when they go they all go together. In an extraordinary mass migration, hundreds of thousands of crabs emerge from the forest. They clamber over everything in their path – including gardens, railway tracks, roads and golf courses – in a frantic scramble for the sea. Close to the shore the male crabs dig out mating burrows, into which they entice females. After the eggs are fertilised, the females go to the water's edge and shake out the eggs. The crabs are so adapted to a life on land that they drown if they are washed out to sea by a wave. The eggs, meanwhile, are taken out with the tide, where they drift with other planktonic organisms. They develop into various larval stages, and every four or five years the ocean currents wash them back onto the Christmas Island shores. Here they metamorphose into a final larval stage – known as megalops, literally 'big eyed', because of the large eyes they develop – and are able to climb out of the sea and head back to the forest.

Hundreds of millions of crab larvae start out on the journey, but not all finish it. Cannibalistic adult crabs intercept some. They stand in the middle of the streams of returning larvae, using their large claws to shovel them into their mouths. The survivors continue into the forest, where they hide below boulders and complete their metamorphosis into miniature adult crabs.

DRAGONS AND SERPENTS

The largest lizards on earth live on volcanic islands farther east than Christmas Island. The longest is Salvadori's monitor (*Varanus salvadori*) on New Guinea, which has been known to grow up to 10 ft 7 in (3.2 m) long. Not quite so long, but heavier and otherwise larger, is the Komodo dragon (*Varanus komodoensis*), which inhabits the Indonesian volcanic islands of Komodo, Flores, Gili

DIPPING CRABS *Red land crabs, having left the rain forest to spawn in the sea, gather around a blowhole on the rocky shore of Christmas Island in order to keep their gills moist.*

AN EXPLOSIVE ISLAND ARC

Some of the most violent volcanoes in the world are found in Indonesia. In this geologically complex region, volcanic islands have appeared and disappeared in dramatic eruptions that first destroyed living things and then played host to new colonists. In the volcanically active eastern section of the Indian Ocean, Krakatoa blew its top in 1883 and the shock waves reverberated around the world. All life there was extinguished immediately, but when in 1929 Anak Krakatoa – 'child of Krakatoa' – emerged from the sea, life was ready and waiting to find its way to the new volcano's shores. Today thickets of casuarina and wild sugar cane grow on its flanks. On a remnant island of the first volcano, the once bare slopes are now covered by tropical forest, some of the seeds having floated here across the sea. Others were carried by wind or brought on the feet or in the stomachs of birds. In the forest live winged creatures – birds, bats and insects – which clearly had no trouble in flying to the island, but there are also pythons, rats and monitor lizards that must have floated there on rafts of driftwood or vegetation that were washed down tropical rivers into the sea. It is all living proof of the way in which volcanic islands provide animals and plants with new opportunities to build communities.

BORN-AGAIN VOLCANO *Anak Krakatoa ('child of Krakatoa') erupts again.*

Mota, Owadi Sami, Rintja and Padar. At one time in the ancient past the 'dragons', which can be up to 10 ft (3 m) long and weigh about 300 lb (135 kg), fed on herds of dwarf elephants, but today they feast on domestic or feral animals brought to the islands by people. They are able to swim from island to island.

These two giants are relatively scarce and live in only a handful of places. The world's largest reptile, by contrast, is widespread along coasts throughout tropical Asia and the Pacific. Indeed, the saltwater crocodile (*Crocodylus porosus*), which can grow to as much as 30 ft (9 m) long, is the most widespread of all the world's crocodiles. It is responsible for many terrifying attacks on people, and has even been known to attack boats and swallow small outboard motors. This seagoing crocodile must, however, return to the land to breed. The female 'salty', as she is known in Australia, takes several days to gather together a nest of vegetation mixed with soil, in which she carefully deposits her eggs. About 90 days later, the calls of the hatchlings summon her back to the nest. She helps to dig them out and then carries them to the water in a special pouch in the bottom of her jaw.

On rocky islands throughout South-east Asia another marine reptile, the banded sea krait (*Laticauda laticauda*) – a sea snake – comes ashore, along with hundreds of its kind. It mates in a seething mass of slithering bodies and deposits its eggs deep in crevices between the intertidal rocks. The hatching youngsters crawl directly back to the sea.

Although sea snakes have evolved from cobras and kraits and retain characteristic snake shapes, they have adaptations for life at sea. The end of the tail is flattened like a paddle, and one lung stretches the length of

LIVING DRAGON *For all its bulk, the Komodo dragon, the largest lizard in the world, is a nimble runner. Some individuals live to be as much as 100 years old.*

GIANT CROCODILE *The largest living reptile is the estuarine crocodile (right), whose eggs are deposited in a nest of vegetation (above), often on coasts of the islands of South-east Asia and of north-east Australia.*

the body, from the head almost to the tip of the tail. This provides oxygen and helps to control buoyancy. Sea snakes can also absorb oxygen – about 30 per cent of their needs – directly from the water through their skins. This enables them to remain below the surface for two to three hours at a time.

Sea snakes take in salts through the skin and ingest them with food; they discharge salts from a gland under the tongue – other marine reptiles such as sea turtles have similar salt glands. The tongue itself 'tastes' the seawater and detects prey. The snake probes cracks and crevices until it finds one inhabited by a fish. It then curls its body against the hole to prevent the prey escaping, and pushes in its head, grabbing the fish in its mouth. Hollow poison fangs, set towards the back of the mouth, inject a poison three or four times more toxic than that of the most poisonous terrestrial snakes. It paralyses muscles and nerves, enabling the snake to swallow its prey with the minimum of fuss.

GULF OF CALIFORNIA

Spiders are rarely associated with the maritime environment, yet on islands in the Gulf of California they have become marine predators. The spiders in question are *Metepeira arizonica* and *Argiope argentata*, and they live on a group of small, barren islands centred on Angel de la Guarda, where they spin webs attached to cacti and rocks on the seashore. Surprisingly, there are many more spiders on the islands than there are on the lushly vegetated mainland nearby. There are sometimes as many as 1000 individuals on a single 6ft (1.8m) high cactus. So where, wondered scientists from Nashville's Vanderbilt University, are the spiders finding their food? The unexpected answer was that it comes from the sea. The spiders feast on kelp flies (which feed on decaying algae) and on other insects that scavenge on carcasses from the islands' huge sea-bird colonies.

Elsewhere in the gulf are former islands that no longer reach the surface of the sea. These seamounts play host every day to a wildlife spectacular involving scalloped hammerhead sharks (*Sphyrna lewini*). Large schools, each containing up to 100 sharks, emerge from the gloom and follow the contours of the seamount. If the school passes through a shoal of fish, the sharks ignore the prey. They just keep moving, swimming endlessly back and forth. As the sharks swim in formation, one will occasionally perform a 'shimmy dance', rotating through 180° and sometimes nipping its neighbour. What this display signifies is unknown, but the robotic movement of the school as a whole is known as 'refuging' behaviour. The hammerheads have switched into automatic pilot, and are simply resting – even for sharks there is safety in numbers. At dusk the school breaks up, and each shark goes its own way to hunt during the night. At dawn all the sharks return.

ISLAND SPIDERS *Silver argiope spiders thrive on kelp flies which scavenge on sea-bird carcasses.*

Coral Reefs, Islands and Atolls

A single square yard of coral reef is home to several hundred thousand organisms. From algae to seaweeds, gaudily coloured fishes and crabs to the coral polyps themselves, they make the reef as rich, size for size, as any tropical rain forest.

Taken together, the world's coral reefs are, without doubt, the largest and most spectacular structure created by living things. They are the result of a simple but significant relationship between a plant – a kind of algae – and an animal: a minute creature known as a polyp. The reefs' surfaces are growing constantly as the polyps secrete a limestone skeleton at a rate of about 1/2 in (1.3 cm) a year. Over thousands or even millions of years, coral deposits build into massive calcareous reefs which can be hundreds of feet thick. Entire mountains in some parts of the world, such as China and Australia, are composed of coral reefs laid down 250 million years ago from which the oceans have since receded.

Sea and Sand *Gradually plants and animals colonise cays such as Bird Island here in the Seychelles.*

Today's living reefs are much younger, dating from the end of the last Ice Age about 10 000 years ago. As sea levels started to rise, the ocean flooded coastal shelves and drowned older reefs. New reefs developed on these earlier foundations, their growth keeping pace with any further rises in sea level. They occur off mainland coasts and around islands set in tropical seas between about 30°N and 30°S. The largest is the Great Barrier Reef off Australia's eastern coast. It consists of 2100 individual reefs and 540 islands with fringing reefs, and extends along 1430 miles (2300 km) of coastline. In the Caribbean the coastal waters off eastern Honduras, Belize and Nicaragua also boast impressive coral reef systems.

Coral Cay Build-up

Coral islands develop in a variety of ways. Atolls – consisting of coral islets and barrier reefs enclosing a lagoon – are usually formed on the top of an extinct volcano. Other coral islands, such as the tiny cays of the Great Barrier Reef, are made up entirely of the skeletons of coral and other calcium-rich reef organisms. They start as barren platforms that accumulate fragments of broken coral and shells, until the sea pushes up dunes high enough to protrude above the water at high tide. A few organisms start to colonise them, though they have to be hardy to do so since there is no shade and little rainwater.

On some coral islands in the Great Barrier Reef it is the animals that arrive first. Scavengers are often the pioneer species, including the seaside earwig (*Anisolabis maritima*) and a small gnat, *Loptocera fittkani*. These feed mainly on guano (accumulated bird droppings) and carrion; the earwig devours this food from below, the gnat from above. Other insects, such as tenebrionid beetles (*Gonocephalum*) and the cricket *Telegryllus oceanicus*, together with predators, such as wolf spiders (*Lycosa*) and giant centipedes (*Scolopendra*), raft in on driftwood.

The first plants germinate either from wind-blown seeds or from seeds drifting in the ocean currents. The plants must be salt-tolerant, drought-resistant, able to cope with a shifting substrate and capable of rapid reproduction. Swine-cress (*Coronopus integrifolius*) copes with these demanding conditions by germinating in winter from seeds dispersed the previous year. It flowers and fruits rapidly and then dies. The rapid dispersal of the seeds ensures that some will sprout the following year. Beach morning glory (*Ipomoea pescaprae*) is another pioneer plant species, sending out runners that bind the sand and prevent erosion.

A lattice of plants eventually covers the island, preparing the ground for other species of both plants and animals. Mites

Sand Cay Formation *An aerial photograph of the Great Barrier Reef shows the early stages of coral island formation.*

CORAL-REEF PIONEERS ***Amongst the first plants to colonise coral reefs is morning glory (above), which is supplied with fertiliser from the guano of sea birds like the black or white-capped noddy (right).***

and small spiders parachute in on the wind, using thin gossamer threads of silk, while lizards, such as the sand goanna (*Varanus gouldi*), and snakes, such as the blind snake (*Typhlina bramina*), arrive on large pieces of driftwood or floating vegetation.

Once the new coral cay has been stabilised, still more plants arrive. Some seeds are carried in the guts of birds and released in their droppings. Others cling or stick to feathers and legs. The fruits of the chaff flower (*Achyranthes aspera*) have hooks, while those of the tar-vine (*Boerhavis diffusa*) are covered with a sticky gum. By this time the island will have divided into two main zones – the beach, with its tough pioneer species, and the hinterland with less hardy species.

The next stage is the formation of a ring of shrubs, which prevents the salt spray from entering the central zone, allowing many more species of plant to take hold, including trees. The shrub ring will probably include the octopus bush (*Argusia argentea*), whose seeds may drift for thousands of miles across the ocean before being washed up on a beach, and sea lettuce trees (*Scaevola sericea*).

A common tree of the Great Barrier Reef is the pisonia (*Pisonia grandis*). It requires plenty of phosphates and nitrates to grow and gets these from the guano of black noddies (*Anous minutus*). These graceful terns nest in the branches, while short-tailed shearwaters (muttonbirds, *Puffinus tenuirostris*) excavate their nest burrows in the ground below.

CLOSE RELATIONSHIPS

Incredibly, coral reefs are manufactured by two of the lowliest organisms on Earth. Coral polyps are tiny creatures with cylindrical stone-like external skeletons. These are attached at one end to a rock or another polyp; at the other end they are open. Tentacles protruding from the open end enable the coral to pull food into its mouth.

By themselves the corals could never build a reef, because they require something which they alone cannot manufacture or acquire in large enough quantities: energy. The sole limitless source of energy is the sun, which can be harnessed only by organisms that are capable of photosynthesis. To achieve their feats of building, corals need a special kind of algae known as zooxanthellae that live in small cavities in the corals' own tissues. During photosynthesis

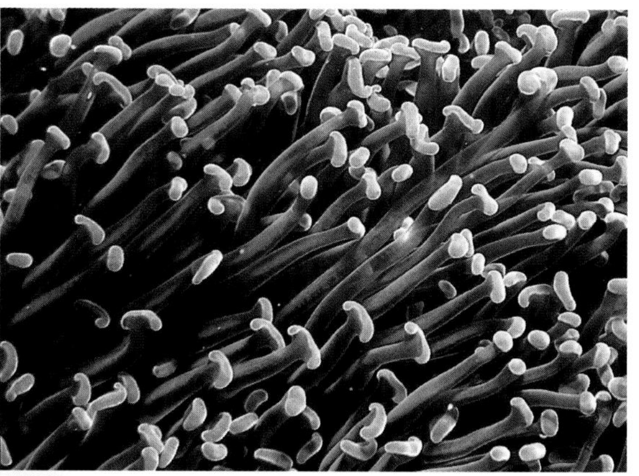

COLOURS OF CORAL ***Stony coral from Indonesian reefs (below) and soft coral from off Norfolk Island (right) demonstrate some of the vivid hues of living coral.***

BIG SHARK, TINY PREY

The largest fish in the world is the 40 ft (12 m) long whale shark (*Rhincodon typus*). But, like the great baleen whales, it feeds on some of the smallest creatures in the sea. Most of the whale shark's life is spent in deep tropical waters, but from time to time it ventures inshore to take advantage of a superabundance of food, the result of events that have taken place on coral reefs.

One regular annual haunt of whale sharks is the Ningaloo Reef, off Western Australia. An estimated 200 of them arrive at the same time each year, during the months of March and April. They appear about a week to ten days after all the corals of the Ningaloo Reef have spawned. The corals of the reefs off Western Australia, unlike those of the Great Barrier Reef on the other side of the island continent, spawn in autumn.

The whale sharks do not eat the coral spawn but wait for a sudden boost in the growth of zooplankton – which, in their turn, take advantage of the availability of protein in the water due to the coral spawn. The sharks follow the vertical migration of plankton, feeding lower down by day and at the surface during the night.

Their tiny prey include the larvae of mantis shrimps and crabs, arrow-worms, comb jellies and copepods. During the weeks after coral spawning they have also been observed feeding on swarms of a tropical species of shrimp-like krill, about a third the size of the Antarctic krill, that during the day spawns in great swarms at the surface. The whale sharks swim through the swarms with mouths agape – many of the huge fish feed at the same time on a flat, calm sea.

Filter Feeding *Mouth agape, a huge whale shark vacuums up plankton at Ningaloo Reef.*

the algae are able to 'fix' carbon from the seawater, transferring much of it to the coral and thereby helping it to grow. The algae also help the coral to take up nitrogen and other inorganic nutrients.

The accumulation of nutrients, and their concentration in one place, benefit algae and coral alike. Unlike temperate and polar seas, tropical seas are generally low in nutrients and therefore deficient in plankton – they are, in effect, nutrient deserts. The internal environment of the coral, by contrast, is relatively rich in vital minerals and chemicals. Phosphates are 1000 times higher in polyp tissues than in the surrounding waters, and ammonium (from which nitrogen is derived) is 5 to 50 times higher. Moreover, the supply of these chemicals is relatively stable inside the polyp, whereas outside it fluctuates with changing sea conditions.

Reef Communities

The kinds of coral that form reefs are known as 'stony' corals. It is their stone-like skeletons that form the reef's building blocks. There are, however, other kinds, known as 'soft' corals, which have larger polyps and do not rely on algae to help them to feed. They have internal rather than external skeletons consisting of needle-like calcareous 'spicules'.

Whether of the hard, reef-forming kind or the soft kind, corals grow into all sorts of

Destruction *Algae growing on a coral reef in the Red Sea indicate the influence of fertilisers and other pollutants, and the probable decline of reef organisms.*

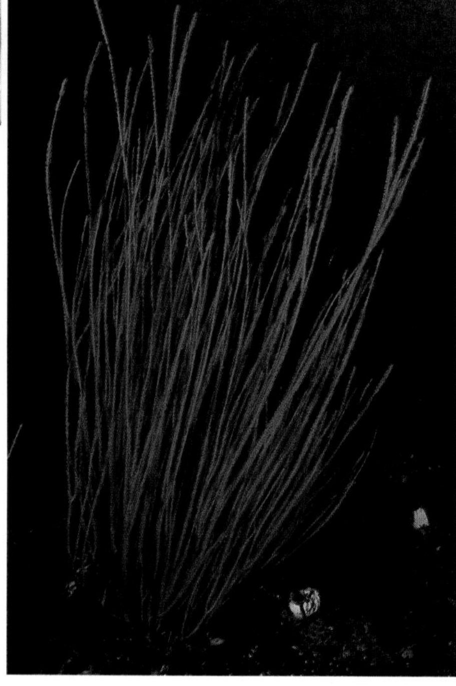

shapes and sizes. There are large tightly packed brain and mushroom corals, branching staghorn and needle corals, pink-coloured daisy corals, leathery soft corals, black corals, Gorgonian fans and whip corals, organ-pipe corals, table, plate and button corals, and fire, bead and bubble corals. Cementing them all together are pink mats of algae.

The result is an environment filled with a diversity of ecosystems, each providing living space for a surprising assemblage of animals. These include coral gall crabs, such as *Happalocarcinus marsupialis*, that hide in galls (outgrowths) on the branches of coral heads. There are decorator crabs (*Camposcia retusa*) which cover their bodies with fragments of seaweeds, sponges and other debris, while sponge crabs (*Dromiopsis edwardsi*) carry entire sponges on their backs. Hermit crabs (*Paguritta*) live in 4 in (10 cm) long tubes in large corals. Snapping shrimps (*Alpheus*) can kill small fish with a single snap of their enlarged claws.

Fish come in a range of magnificent colours, each species with its own niche and its own peculiar way of going about its business. The striped catfish (*Plotosus lineatus*) forages in tight 'balls' several yards across, while the spindly trumpet fish (*Aulostoma chinensis*) shadows the obese puffer fish (*Arothron stellatus*) so closely that it follows its every move. Scorpion fish (*Scorpaenopsis*) disguise themselves as seaweed-encrusted rocks; stonefish (*Synanceia horrida*) resemble the rock itself. Three-spotted damselfish (*Dascyllus trimaculatus*) share the crowns of sea anemones with clown fishes (anemone fishes, *Amphiprion*). They acquire an immunity to the stinging cells on the anemone's tentacles by covering themselves in a layer of mucus from the anemone itself. They do this gradually, swiping their tails first across the tentacles, then other parts of the body, until the vital coat has built up. When the fish brushes against the tentacles, the anemone probably assumes that it has touched itself and does not activate its stinging cells.

WEALTH OF THE REEF ***From closely packed mushroom corals (above) to grass-like whip corals (left), the reef encompasses a huge variety.***

Shoal Searching *A shoal of striped catfish, moving almost as one in dense shoals safe from predators, scours the sea floor for food.*

In the evening the animals of the dayshift retire, and a different set of creatures emerges for the night. Dusk is a particularly frantic time. White-tipped reef sharks (*Triaenodon obesus*), which have been resting on the reef floor by day, suddenly become active, chasing prey among the nooks and crevices between the coral heads. Coral polyps emerge from their hard cases. The loss of light means a loss of photosynthesis for those corals with symbiotic algae, so they fish for plankton instead.

Feather stars (*Comanthus*) emerge from their hiding places and find suitable vantage points where they can wave their spindly arms in the current like underwater lilies. Their multibranching relatives the basket stars curl up tight at the slightest hint of danger. Sea urchins, such as the Red Sea's *Asthenosoma varium* with its armoury of spines and poison sacs, graze the algal mats, their nocturnal wandering reducing competition from other herbivores, such as parrotfish (*Scarus*), that are active by day. The parrotfish themselves sleep in mucous envelopes to lessen the ability of predators to smell their presence. Grunts (family Pomadasyidae) mute their brightly striped body colours, essential for keeping together in schools by day, and take on a blotchy hue that camouflages them against the coral.

Reef Residents *The legs of red hermit crabs protrude from their shelters in a Caribbean reef.*

Cleaning Services

The health of a coral reef's entire fish community is dependent on the specialist services of two types of diminutive reef resident: cleaner fish and cleaner shrimps. Like barbers touting for customers, they wait at well-established cleaning stations where large fish, such as groupers, sharks and manta rays, and smaller fish, such as batfish (*Platax orbicularis*), come to divest themselves of irritating parasites.

The cleaner wrasse (*Labroides dimidiatus*) of the Great Barrier Reef has a blue stripe along its silvery-white body, nature's equivalent of the barber's pole, to advertise that it is open for business. When a fish arrives to be cleaned it hovers over the cleaning station, and the wrasse performs a little dance. This reassures its customer that it is a legitimate cleaner fish; fish rarely allow others to come too close. The cleaner then inspects the customer's body, and begins its work.

Dangerous Liaison *Secure inside a layer of protective mucus, clown fish find shelter among the stinging tentacles of a sea anemone.*

CLEANING STATION *A coral trout is cleared of its parasites by a group of striped cleaner wrasses which work out of a recognised cleaning station.*

The larger fish allow the cleaners to enter even their mouths and gill chambers without the smaller fish coming to any harm.

Chemical warfare plays a part in reef life. Mushroom corals, such as the Hawaiian *Fungia scutaria*, are stand-alone corals which occupy bare patches of sand in an otherwise overcrowded reef. Most of the space on coral reefs is taken up by reef-building corals, but mushroom corals have a way of making room for themselves. Like other corals they produce mucus, but theirs contains stinging cells, known as nematocysts (the same stinging cells as those in the tentacles of sea anemones and jellyfish), which attack the polyps of other corals. In this way the mushroom coral avoids becoming overwhelmed.

POISONOUS PREDATOR *The blue-ringed octopus of Australian coral reefs is one of the most venomous creatures in the sea.*

Chemical weapons are used extensively by another group of reef inhabitants: the sea slugs (nudibranchs). Some have acid-secreting cells, while others appropriate the stinging cells of their sea anemone prey and store them in special finger-like sacs on their backs. They display their unpalatability by wearing bright colours, such as those of the Spanish dancer (*Hexabranchus sanguineus*) in the reefs of the Indian and Pacific oceans; it has a bright scarlet-and-orange body and a red-and-white striped fringe to its mantle. Among sea slugs, the Spanish dancer is remarkable for obtaining a particularly potent poison, halichondrimide, from bread-crumb sponges (*Halichondria*). Not satisfied with halichondrimide's potency, it modifies it into two poisons, known as macrolides, that are the most effective substances known for repelling predatory fish. The Spanish dancer uses the same chemicals to protects its eggs – conspicuous pink ribbons which it leaves coiled around coral debris. The only difference is that the poison used to protect the eggs has ten times the concentration of that found in the adult.

GAUDY GIANT

A monster sea slug lives off the Seven Brothers Islands in the narrow mouth of the Red Sea near Djibouti. It can grow more than 20 in (50 cm) long and 14 in (35 cm) wide, and weigh as much as 4 lb 6 oz (2 kg). It is brilliant pink and peach in colour, has a bubbly texture to the dorsal skin, and sports a pair of yellow tentacles and a rosette of feathery gills 8 in (20 cm) across. It is thought to be the largest sea slug in the world and was discovered as recently as 1991. It may be an unusually large variety of the Spanish dancer, or it may be yet another species of *Hexabranchus*, entirely new to science.

The blue-ringed octopus (*Hapalochlaena maculosa*) of Australian reefs is another careful, chemical-conscious parent. The adult octopus is well known to be a highly venomous creature. It bites its shellfish prey with its horny beak and injects a lethal nerve poison, known as tetrodotoxin, with a toothed tongue-like radula. The chemical is powerful enough to kill people, too. The surprise is that the octopus also provides its eggs with an identical poison that serves to protect them against egg predators.

Taking care of eggs is important for any animal in the reef community, since there are always egg predators ready to scoop up

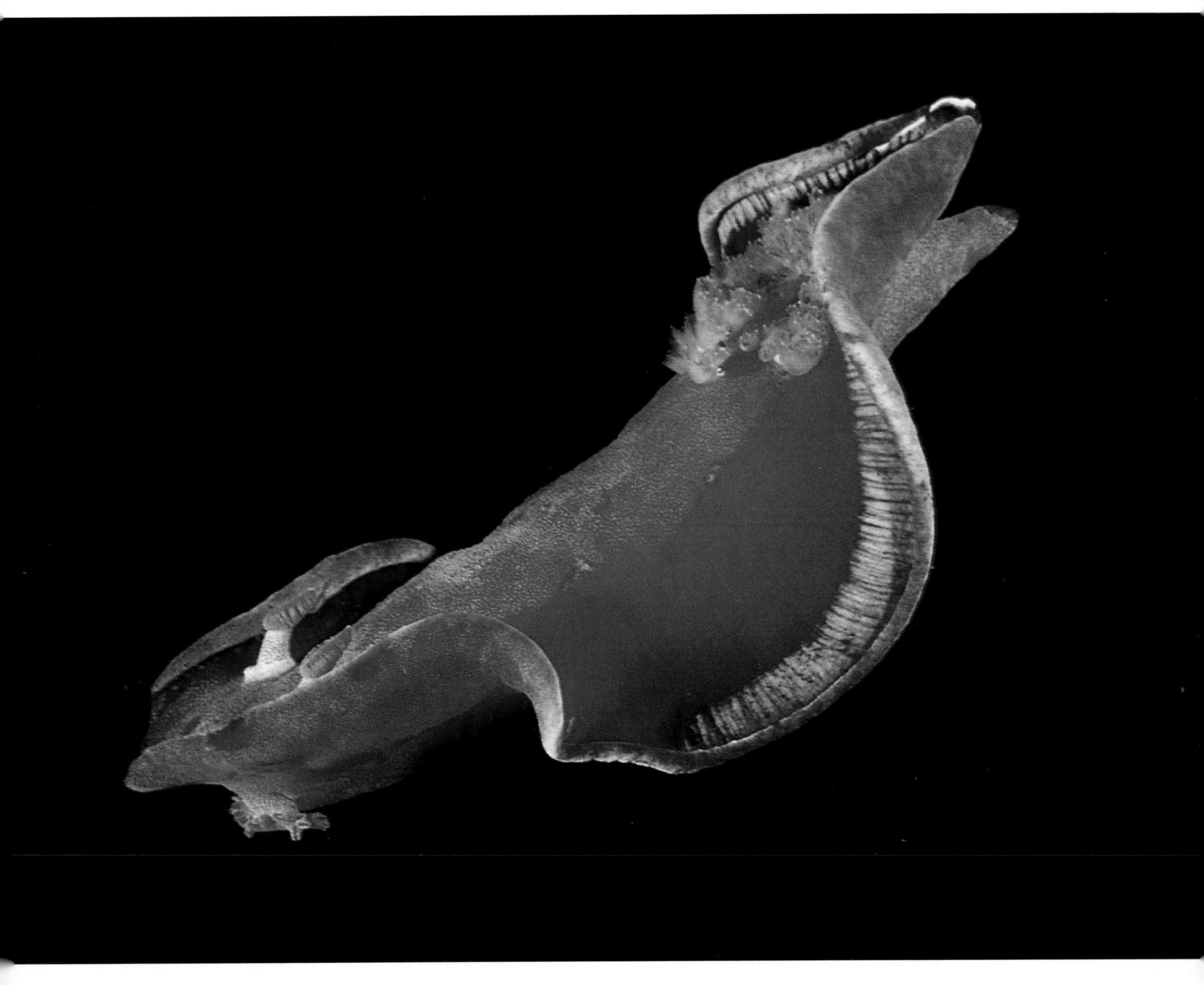

WARNING SIGNS *The Spanish dancer of the Great Barrier Reef has a bright livery which acts as a warning to potential predators that it is dangerous to eat.*

an unprotected brood. Among the clown fish (*Amphriprion clarkii*) of the Indian and Pacific oceans this is the task of the male. Clown fish usually live in pairs that remain together for a couple of breeding seasons. They produce between five and seven clutches during the summer, the male guarding the eggs he has fertilised. He takes away any dead or infertile eggs, and fans the rest with his pectoral fins, so that they receive enough oxygen for development. The female plays no role in the brooding process unless the male falls prey to a larger fish, when she takes over until another male stops by.

Like damselfish, clown fishes protect themselves by living among the tentacles of large sea anemones. But life here is not as safe as it might appear. Males can be displaced by other males, but strangely enough, when a new one joins a female with eggs fertilised by another he does not destroy them as most male animals would do. Instead he guards his predecessor's eggs. Not that he is a willing foster father: it is the female who bullies him into taking on his new role. The female clown fish is larger than the male, and she butts him into submission with a powerful blow to his belly. The male shows deference by standing on his head and biting the rock next to the eggs. His patience and diligence pay off, for within 20 days of the eggs hatching he is guarding a clutch of eggs that he has fertilised himself.

The male Cortez damselfish (*Stegastes rectifraenum*) from the Mexican coast is left guarding the eggs, too, but this father is not so particular. He eats about 5 per cent of the eggs. Experts believe that eating a few eggs now and then enables a male to grow bigger, survive the current breeding season

REEFWATCH *Like its cousin the Cortez damselfish, the male three-spot damsel (*Dascyllus trimculatus*) is left to guard the eggs.*

and father even more eggs next time, thereby increasing his reproductive success in the long term.

SPAWNING CULTURE

Another way of ensuring that a new generation will be there to take over from the present one is to produce so many eggs that predators are quickly sated, allowing some offspring to survive. Many of the reef's invertebrates achieve mating success in this way, with all members of the species spawning at the same time. The process is usually synchronised by the phases of the moon, though solar events, such as the solstices, are also important. For example, corals around South Pacific islands all spawn simultaneously when the moon enters its third quarter each October and November. In the same reefs the burrowing worms known as palolo worms (*Palola* and *Eunice*) discard their egg and sperm-filled tails at dawn just two days later in another example of synchronised spawning. Equally astonishing is the event that takes place on the Great Barrier Reef during the Southern Hemisphere's spring or early summer. About 130 species of hard corals along the entire length of the reef spawn over a series of nights after a full moon. And they are not

UNDERWATER SNOWSTORM *Most of the corals on the Great Barrier Reef spawn (bottom) during a few nights shortly after a full moon. The eggs and sperm combine and then divide to form tiny floating embryos (inset).*

South Pacific Starfish *A red brittle star crawls amongst soft corals on a Marshall Islands reef.*

alone: sea cucumbers, sea urchins, starfish, feather stars and sponges all spawn together at the same time as the corals.

For the Great Barrier Reef's crown-of-thorns starfish (*Acanthaster planci*) – the corals' arch enemy, because it eats the polyps – it is a chemical in the water that prompts it to switch on and spawn. The chemical is present in sufficient quantities when the population of starfish is large enough. Then all the starfish, each of which grow up to 28 in (70 cm) across, spawn together. If the population is too small, the starfish fail to spawn.

This interdependence of animals and plants in the reef is also highlighted by the Caribbean sea urchin (*Diadema antillarum*) and the brown-green algal turf on which it feeds. These long-spined black sea urchins graze the algae, but where the urchins are absent the turf grows less well. It turns out that the waste products from the sea urchins, in particular ammonium containing vital nitrogen, are just the chemicals needed by the algal cells to grow, so the two feed and grow in some kind of harmony.

A starfish from the Caribbean poses no threat to its host reefs, as the crown-of-thorns starfish does; instead, it has to protect itself from predators and does so in a rather surprising fashion. By day the brittle star, *Ophiopsila riisei*, hides away in cracks and crannies in the reef, but at night it allows about three-quarters of its body to emerge, to feed on suspended matter in the water. At this time it is vulnerable to attack by reef creatures such as predatory swimming crabs (*Portunus*). If touched, it first gives off faint warning flashes of light; then, if the threat persists, it produces waves of light that travel from the centre of its body to the tips of its arms. The flashing not only startles the predator but also warns it that the brittle star is unpleasant to eat.

Reef Colours

Colour is evident everywhere on a coral reef, although how many reef creatures can appreciate the subtle hues, tints and pigments is unknown. One group of animals that certainly can, and to a remarkably sophisticated degree, is the order Stomatopoda – the mantis shrimps.

Mantis shrimps are some of the most vicious crustaceans known, and they hole up in coral reefs. Armed with clubs, spears and body armour, these extraordinary creatures do battle like ancient knights. Their weapons are so powerful that they can pierce a human hand with a spike, or crack a fingernail with a blow that has the velocity of a rifle bullet. To enable it to deploy such elaborate weaponry the mantis shrimp possesses an equally sophisticated sensory system.

It has a pair of stalked eyes, each of which is divided into three sections. The top and bottom hemispheres resemble an insect's compound eye, filled with rows of tiny lenses called ommatidia; in between is a horizontal band containing much larger ommatidia. These have brightly coloured filters through which light must pass before hitting the light-sensitive cells at the back of the eye. Different species have different filters, possibly implying that each species of mantis shrimp sees different colours. The central band also has two rows of receptors that are thought to be able to detect the plane of polarisation of light.

Both eyes move independently, like those of a chameleon, the upper and lower hemispheres and the central band in each eye scanning the same area of space. By comparing the image gained from the hemispheres with that from the horizontal band, the mantis shrimp has a useful rangefinder. With this advanced visual system, the shrimp is able to analyse the location, colour and polarising properties of a target, making it a remarkably accurate predator.

Eyes on Stalks *The eyes of the mantis shrimp are able to see colours and accurately locate prey.*

Islands of the Southern Ocean

In the storm-tossed Southern Ocean tiny specks of land provide summer homes and breeding grounds for millions of sea birds and seals. They are all here because of a minute shrimp-like creature just over 2 in (5 cm) long – the krill.

A ring of windswept isolated islands is the focus for a wealth of wildlife in the high latitudes of the Southern Hemisphere. Some volcanic in origin, others formed from uplifted sedimentary rocks or ocean crust, they are scattered in a band between latitudes 46° and 59°S. They dot the enormous gap left by the break-up, starting more than 150 million years ago, of the southern supercontinent Gondwanaland. Deep trenches opened up between what are now Australia and Antarctica, the Antarctic Peninsula and South America, creating an ocean which is some 26 000 ft (8000 m) deep in places. Sandwiched between the waters of the South Atlantic, South Pacific and southern Indian Ocean to the north and those of the Antarctic to the south, the currents of the Southern Ocean sweep eastwards as they circle the bottom of the globe.

LUNGE-FEEDING WHALES *In the summer, humpback whales migrate from the tropics to the Southern Ocean where they feed on krill.*

A handful of subantarctic islands or island groups are the only tiny parcels of land to break the ocean's surface. They consist of South Georgia, South Sandwich and Bouvet, to the south of the Atlantic Ocean; Prince Edward and Marion Islands, Crozet, Kerguelen, Heard and MacDonald Islands, to the south of the Indian Ocean; and Macquarie Island, to the south of the Pacific Ocean. As remote from each other as they are from the continents, they straddle the area, known as the Antarctic Convergence Zone, where the warmer waters from the north meet the cold waters from farther south. This is a zone of upwelling where nutrient-rich waters are brought to the surface, forming the basis of a food chain that ends in the largest animals on the planet, including the very largest of all, the blue whale (*Balaenoptera musculus*).

Isolated Islands

Conditions on the islands are rugged. Only a fairly small range of plants live there, including just 24 species of grass, 32 of herbaceous plant and 16 of fern in all the islands put together. Other species, however, are better adapted to the harsh conditions and are represented in greater numbers. There are more than 350 species of moss, 300 of lichen, 150 of liverwort and 70 of mushrooms and toadstools. Islands to the south of the Antarctic Convergence Zone – South Georgia, South Sandwich and Bouvet – tend to be glaciated and snow-covered; the vegetation consists mainly of mosses, lichens and a few flowering plants. Those on the convergence or to its north – Crozet,

SEAL ROOKERY *As densely packed as human sunbathers at a popular resort, Antarctic fur seals gather on a South Georgia beach to breed.*

UBIQUITOUS KRILL *Tiny shrimp-like krill (above right) congregate in vast numbers on the surface of the Southern Ocean (above).*

Kerguelen, Macquarie, Prince Edward and Marion Islands – are free of snow for most of the year. They are green with tussock grasses and herbaceous plants, although there are no trees.

The permanent animal life on the islands includes nematode worms, molluscs, spiders and various insects, but there are no amphibians, reptiles or freshwater fish. Only seven species of bird are permanent residents, including two species of sheathbill, two of duck and a single one of pipit. The only resident mammals, such as reindeer (*Rangifer tarandus*) on South Georgia, have been introduced by humans. But while permanent residents are rare, summer visitors abound. They include birds – penguins, albatrosses, petrels, skuas, shags, gulls and terns – and seals, such as elephant and fur seals. When these birds and mammals arrive, they come in extraordinarily large numbers. On one island alone – Zavodovski, in the South Sandwich group – an estimated 18 to 21 million penguins breed each year. A single beach in South Georgia is the breeding site for 2 million Antarctic fur seals (*Arctocephalus gazella*).

Less numerous species are also present, such as the scarce subantarctic fur seal (*Arctocephalus tropicalis*). Like most other southern seals it was brought to the brink of extinction in the 19th century by commercial sealing; today, however, it is found on several islands, as is another former victim of the sealers, the southern elephant seal (*Mirounga leonina*), whose numbers have recovered to more than 750 000. The elephant seal, the largest animal to land on the subantarctic islands, likes to breed on wide, open beaches. These it finds on South Georgia, elephant seal capital of the Southern Hemisphere. Every spring the huge males – the largest individuals measuring 16 ft (5 m) long and weighing almost 2½ tons – arrive on the beaches and fight for the right to mate. Each victor will dominate a different beach, or section of a beach, upon which the females will haul out a couple of weeks later.

BEACHMASTER CONTEST *Two bull elephant seals fight for the right to mate with a harem of females.*

Offshore, meanwhile, are whales, also recovering from overexploitation in the first half of the 20th century. They include blue, sperm (*Physeter catodon*), minke (*Balaenoptera acutorostrata*) and humpback (*Megaptera novaeangliae*) whales.

KRILL SOUP

All these visitors – whales, seals and sea birds alike – are totally dependent for their survival on one small animal: the ubiquitous krill. There are about ten species of

Stay Cool *Southern elephant seals on a Crozet Island beach throw sand over their backs to keep cool.*

krill in the world's oceans, and the Antarctic krill (*Euphausia superba*) is one of the largest. Even so, measuring only about $2^1/_2$ in (6 cm) long and weighing little more than $^1/_{30}$ oz (1 g), a single krill is, literally, a speck in the ocean. When many krill get together, however, they can form swarms that measure miles across. These may have as many as 49 million individuals per square mile (19 million per square kilometre), and weigh tens or even hundreds of tons. Occasionally 'superswarms' form, covering as much as 174 sq miles (450 km²) of sea and weighing up to 2 million tons. They stain the water red by day and light it up with their eerie blue-green luminescence at night. Worldwide, krill are so numerous that it has been estimated that they weigh more than the world's entire human population.

Krill are a kind of zooplankton (animal as opposed to plant plankton), but they differ from other planktonic organisms in being heavier than water. If they stop swimming, they sink. They move forwards at an angle of about 55°, using five pairs of paddle-shaped legs, and can propel themselves backwards for short distances using their tails. By changing the angle of their bodies in the water they can move up or down.

Light appears to play an important role in the life of krill. Their large black compound eyes are stalked and respond not only to light from the sky above but also to light from the sea below – the glow of other krill. They have luminescent organs on their eyes, on one of their pairs of legs and on the underside of the abdomen. These can be lit up for several minutes or flashed on and off. Light signalling is probably a way in which members of a swarm keep together.

By day, krill tend to remain in deeper water, although they cannot go down any farther than 820 ft (250 m) because they need high levels of oxygen. At night they rise to the surface to feed. They graze not only on phytoplankton – plant plankton – but also on small floating animals such as pteropods (sea butterflies) and even on others of their own kind. They fish for these creatures using a kind of net or basket, made from appendages of their own bodies, consisting of feather-like structures lined with tiny bristles. Plankton tend to concentrate in places where there are upwellings of nutrients, so large aggregations of krill are found there, too.

Krill breed in the southern summer. During the female's moult the male transfers two ball-shaped packs of sperm to a cup-like depression on her underside. The female stores the sperm in a special chamber until the eggs are ready. She then carries her pink egg-bundle below her abdomen until it is time to spawn. As each egg passes a slit in the sperm storage chamber it is fertilised, and as many as 2000 eggs may be released in a couple of hours, a process that is repeated twice or more during the summer. The eggs sink down to depths of more than 2460 ft (750 m), but as the larvae develop over the next couple of years they gradually rise to the surface waters.

For such tiny creatures they have an extraordinarily long life expectation; individuals in captivity have lived for up to seven years. In the wild, however, few will survive that long, for krill are prey to all sorts of animals, from the smallest bird to the largest whale. They are the reason that all the summer visitors come to the bleak and brutal Southern Ocean in the first place.

Ancient Mariner *A wandering albatross travels over miles of ocean in search of food, such as squid.*

TILL DEATH DO US PART *Grey-headed albatrosses perform a courtship display which will cement a lifelong bond.*

Seals and sea birds, for example, all feed directly on krill or on marine animals such as fish that feed on them. Both seals and birds are tied to the land to breed, so a rookery or nest site adjacent to such a productive ocean is essential for offspring to feed well. Land is in short supply, however, and so the subantarctic islands – the only land for thousands of miles – become the temporary homes to many millions of animals.

Seals alone, of which there are an estimated 33 million in the Southern Ocean, account for 128 million tons of krill annually – more than three times the amount that is consumed by whales.

OCEAN WANDERERS

Not far from the elephant seal rookeries on South Georgia, on the flat areas of tussock grasses behind the beaches or on headlands, an epic life cycle unfolds. Here wandering albatrosses (*Diomedea exulans*) have their breeding colonies. As with the elephant seals, the males arrive first, in early November. They busy themselves with the job of renovating the nests of grass, moss and soil, shaped like truncated cones, that they and their lifelong partners use year after year.

Albatrosses are huge – the largest sea birds in the world. The wings of the biggest species can span as much as 12 ft (3.7 m), so nest sites tend to be in places where the birds have sufficient runway space to land and take off. Nevertheless, after spending a year at sea and in the air, the birds can be undignified: sometimes crash-landing, tumbling among the tufts of tussock grasses that serve to soften the bump.

MATERNAL INSTINCT *A female wandering albatross incubates a single egg on its nest amongst tussock grass on Bird Island, South Georgia, while her partner is at sea finding food.*

In late November and early December the females arrive. Partners locate each other and perform a greeting ceremony involving elaborate bowing, walking and pointing rituals. Having restored the bond between them, they work together to finish the repairs to their nest. Then they mate. A little later the female lays her single egg in the nest, and both parents share incubation duties.

Not all the albatrosses that arrive on the islands will breed, however. Young, unmated birds do not get down to mating immediately. As they will stay with the same partner for the rest of their lives (unless a partner turns out to be unable to breed successfully or is killed at sea), they are very careful in selecting the right candidate. A succession of displays throughout the breeding season is intended to attract and impress. It includes 'bill-circling', in which one of a pair moves its bill rapidly in a half circle around its partner's bill; 'sky-pointing', in which both birds point their bills towards the sky; and the spectacular 'ecstatic display', during which the two birds face one another, with their massive wings outstretched, and point their heads to the sky, calling loudly. Sometimes wandering albatrosses display while floating on the surface of the sea, one bird feeding the other with a gift of regurgitated food. Even when two birds are satisfied with each other's performances, they may not breed for several more years. Each season they go through the same rituals until one day they are ready to mate.

Some birds in the breeding colony may be neither displaying nor mating. Instead they will still be feeding the chicks that hatched last year, since young albatrosses spend an entire year in the nest before they are ready to fledge. A chick hatches in March or April, about ten weeks after the egg was laid. It has a thick down coat to keep it warm, and is constantly brought food by its parents, who take it in turns to go to sea to feed and to guard the nest. Skuas and giant petrels will take albatross chicks, given half a chance, but the parents are very protective.

The chick is fed a regurgitated fish soup by whichever parent has done the foraging. When the chick gets bigger and able to fend off attackers for itself, both parents go fishing, leaving their offspring for three or four days at a time. It sits on the nest throughout the harsh subantarctic winter, fed continuously by the adult birds. The youngster puts on a thick layer of subcutaneous fat that, together with the thick down, serves to keep it warm. At this stage chicks can be heavier than adult birds.

During its first spring, the chick begins to slim down and prepare itself for its maiden flight, which will take place in the southern summer, December or January. It expands its huge wings and, using the wind, practises short flights in which it hovers over the nest. Once competently airborne, it flees the nest, leaving alone or in small groups. It does not return to the breeding colony again for six or seven years, when it is its turn to start the long reproductive cycle.

At sea, albatrosses fly vast distances, riding the winds of the Southern Ocean to reach the places where squid and fish are plentiful. They even sometimes circumnavigate the globe between breeding seasons. On their travels they catch fish and squid up to 13 lb (6 kg) in weight, either by landing on the sea and seizing prey that rises to the surface or by shallow plunge-diving. Fish and squid feed on krill, which are close to the surface at night. This is when the albatrosses pick them off – the hunters becoming the hunted.

Downy Giants *Black-browed albatross chicks on raised nests wait for their parents to return from sea with another helping of food.*

Bird Islands

Several albatross species visit the subantarctic islands each year to breed. Black-browed albatrosses (*Diomedea melanophris*) and grey-headed albatrosses (*D. chrysostoma*) form large colonies like those of wandering albatrosses; the light-mantled sooty albatross (*Phoebetria palpebrata*) nests alone or in small groups. Petrels nest here, too, as do sheathbills, prions and the Antarctic tern. With no four-footed predators around, few birds nest on inaccessible cliffs as they do in the Arctic, although some dig burrows to escape the attentions of larger predatory birds.

Petrels are close relatives of albatrosses, ranging in size from the two albatross-sized

Squabbling Scavenger *A southern giant petrel hovers over its nest in the South Orkneys (above). Two petrels fight (top) for food, often the corpses of penguins or seals.*

giant petrels (*Macronectes*) to the many tiny storm-petrels (family Hydrobatidae), some of which are only sparrow-sized – which makes them the world's smallest sea birds. The southern giant petrel (*Macronectes giganteus*) builds its nest among the tussock grasses of subantarctic islands. Pairs of birds are partners for many seasons, and maybe for life. Both parents care for the single chick, although it may have to defend itself against other giant petrels or skuas. It does this by growling and ruffling its neck feathers to make itself look bigger. If an attacker still fails to go away, the chick ejects from its open bill a stinking regurgitated oil which fouls the attacker's feathers.

By far the most numerous birds in the subantarctic zone, however, are penguins, which make up 80 per cent of its total bird

population. Species that breed here include the king penguin (*Aptenodytes patagonica*), Macaroni penguin (*Eudyptes chrysolophus*), rockhopper penguin (*E. chrysocome*), gentoo penguin (*Pygoscelis papua*) and chinstrap penguin (*P. antarctica*). Like all sea birds, penguins must come to land to breed, and for them, too, the subantarctic islands provide refuges in an icy, stormy ocean. Although superbly adapted to a marine life, penguins are much more clumsy on land and sometimes find it difficult to come ashore. Rockhopper penguins ride the breakers, like a lift up a cliff face, to struggle to their colonies at the top. Once on land they use their wings to help them to balance. Chinstrap and Macaroni penguins hop, clamber, waddle and toboggan over the foreshore, up steep slopes and across rocks and scree. Penguin colonies can be vast, containing many thousands of birds.

Over half of all penguins on subantarctic islands are Macaronis. In spring they can

Southern Commuters ***Rockhopper penguins tread a well-worn pathway, complete with toenail grooves, to their cliff-top rookery.***

Penguin Colonies *More than 75 000 pairs of chinstrap penguins (left) nest along the rim of a volcanic crater at Deception Island. Macaroni penguins (right) climb over a rocky shore on the Crozet Islands.*

turn a barren, sloping hillside on South Georgia into a noisy, restless colony with more than 80 000 nests. The males arrive first and sort out any territorial disputes which, as Macaronis are very aggressive, can be ill-tempered affairs. The females arrive a week later, and after courtship and mating are over each female lays two eggs. The first egg is half the size of the second, and only the larger one hatches – the same happens with rockhopper penguins. Why the female should lay two eggs is a mystery. The smaller egg, however, provides food for sheathbills (*Chionis*) – a welcome change for these scavenging birds from their frequent meals of penguin droppings containing undigested fragments of krill.

Unlike Macaronis, gentoo penguins do not gather in huge colonies. They are also much tidier. Their nests, in particular, are immaculate, made from pebbles and feathers. There may be up to 1700 pebbles and 70 feathers in a single nest, and each pebble or feather is the same size. If pebbles are in short supply, these relatively docile birds go on a pilfering spree, stealing them from the nests of other birds. The thieving is so widespread that a single pebble will probably have been part of every nest in the area at some time during the same breeding season.

A third species, chinstrap penguins, seems to have drawn the short straw for nest sites. They are found in the least accessible places at the tops of steep slopes. On Deception Island, in the South Shetlands, about 100 000 pairs nest at the edge of a volcanic crater. Their daily journey when feeding their chicks is extraordinary. They first have to cross the beach and are then channelled along a narrow gorge. The long winding trail of penguins is very orderly. They keep to the right, incoming traffic moving along one side of the path, outgoing along the other. After crossing a river of glacial meltwater, which could easily wash them downstream, they climb the steep slopes of volcanic ash, and finally disappear into the mists swirling around the top of the island.

Southern Sovereign

The king penguin, the largest of the subantarctic penguins, follows an unusual 10 to 12 month breeding cycle, which means that chicks are being born throughout much of the year, in spring, summer and even autumn. At any one time, therefore, a colony will have in it penguins at all stages, from baby to adolescent to adult. Colonies are enormous – some, including one on South Georgia, containing more than half a million birds – and they are never empty.

The breeding cycle, whether it starts in spring, summer or autumn, involves elaborate courtship displays. A key ritual early on is a synchronised walk: one bird following the movements of the other to such an extent that they march as one. Partners that have reached a more advanced stage lift their heads and point their bills to the sky, holding this strange pose for ten seconds or more. They then shake their heads from side to side for several minutes and make loud trumpeting calls.

In due course the newly paired female lays a single egg and, like the female emperor penguin, immediately passes it to the male, who balances it on his feet. She then

Sky-pointing *King penguins display to one another at a nesting colony in the Falkland Islands.*

goes to sea to feed, leaving the male with the egg safely tucked under a warm flap of abdominal skin. Unlike the emperors, though, the male and female share incubation duties until, after about three weeks, the chick hatches. Then they also share feeding duties, one parent always staying behind to guard the chick. Skuas and giant petrels patrol king penguin colonies, ever alert for an unprotected chick or a corpse – in fact, skuas establish territories that encompass seal or penguin colonies precisely to ensure that their nest sites are close to such good food sources.

As the king penguin chicks grow they put on thick, brown, fluffy down, and eventually gather in crèches. Both parents can then go to sea to gather food for their rapidly growing youngster. During the summer conditions in the colony can be extremely uncomfortable. Adults and chicks alike are adapted for cold conditions, and so on warm summer days the birds overheat. Rows of penguins can be seen lying on their bellies, their flippers in the air, and their beaks wide open, panting. On South Georgia a glacial stream provides a convenient place to stand and cool off.

By the southern autumn, in April, there are chicks at all stages of development everywhere in the colony. The spring birds have grown almost as tall as their parents and have put on substantial quantities of fat, essential for survival during the winter months. Their parents do not return so often during winter, so the chicks rely on their stored fat to carry them through the winter fast. The chicks that hatched in late summer often fail to gain sufficient fat reserves before winter sets in, and they die. To make up for the loss, their parents tend to be the early breeders the following spring. In this curious breeding cycle, king penguins rear two chicks every three years.

FLYING SUBMARINES ***Streamlined king penguins flock under water just as other birds gather in the air.***

INDEX

References in *italics* denote illustrations

Picture Credits

Abbreviations

T = top; C = centre; B = bottom;
R = right; L = left
BCL = Bruce Coleman Ltd
FLPA = Frank Lane Picture Agency
NHPA = Natural History Photographic Agency
OSF = Oxford Scientific Films
SPL = Science Photo Library

3 OSF/Kim Westerskov. **6** OSF/Fredrik Ehrenstr. **7** Colorific/Linda Bartlett. **8** Auscape International/Jamie Plaza Van Roon, TL; Icelandic Press & Photographic Service/Matts Wibe Lund, BL. **9** Icelandic Press & Photographic Service/Matts Wibe Lund, B; FLPA/S. Jonasson, TL. **10** BCL/Pekka Helo, BC; FLPA/E. & D. Hosking, TL. **10-11** BCL/David R. Austen. **12** SPL/Simon Fraser, TL; Planet Earth Pictures/Warren Williams, BL; Colorific/Al Grillo, TR. **13** SPL/Sam Pierson. **14** Planet Earth Pictures/Pete Atkinson, TL; Biofotos/Heather Angel, BC. **15** DRK Photo/Larry Ulrich, TL; OSF/Michael Fogden, BR. **16** OSF/Jack Dermid. **17** Robert Harding Picture Library. **18** Planet Earth Pictures/Space Frontiers, TR; Auscape International/Reg Morrison, BL. **18-19** Siena Artworks Limited/Leslie Smith. **20** Siena Artworks Limited/Leslie Smith, TL; FLPA/H.D. Brandl, CL. **21** Planet Earth Pictures/John Lythgoe. **22** OSF/Michael Brooke, BL; Kevin Jones Associates, TR. **23** Auscape International/Jaime Plaza Van Roon, T; BCL/Atlantida SDF, Inset; Siena Artworks Limited/Leslie Smith, BR. **24** BCL/Franz Lanting, L; FLPA/S. McCutcheon, TR. **25** G.R. Roberts, BL; Siena Artworks Limited/Leslie Smith, TR; OSF/G.I. Bernard, CR. **26** BCL/Jens Rydell, T; Comstock, BL. **27** OSF/John Netherton, CL; BCL/John Shaw, C; DRK Photo/Peter French, TR; BCL/Bob Glover, BL; Siena Artworks Limited/Leslie Smith, BR. **28** BCL/Jan Van De Kam, C; OSF/Stan Osolinski, T. **29** Tom Stack & Associates/NASA/TSADO, T; Siena Artworks Limited/Leslie Smith, BR. **30** Auscape International/D. Parer & E. Parer-Cook. **31** Kevin Jones Associates, T; SPL, C. **32** Kevin Jones Associates, TL; SPL/NASA, TC. **33** NHPA/Peter Parks. **34** NHPA/David Woodfall, TL; NHPA/Anthony Bannister, CR. **34-35** FLPA, B. **35** Siena Artworks Limited/Leslie Smith. **36** DRK Photo/Darrell Gulin, T. **37** DRK Photo/Norbert Wu. **38** OSF/Aldo Brando, BL. **39** Tony Stone Images. **40** Kevin Jones Associates. **41** Auscape International/D. Parer & E. Parer-Cook, TL; Planet Earth Pictures/Keith Scholey, Inset; Planet Earth Pictures/Roger de la Harpe, B. **42** BCL/Dieter & Mary Plage. **43** Kevin Jones Associates, BL; FLPA/E. & D. Hosking, TR. **44** Auscape International/Jean-Paul Ferrero. **45** OSF/Isaac Kehimkar. **46** FLPA/NASA, TL. **46-47** DRK Photo/Larry Lipsky, B. **47** Colorific/J.B. Diederich, CR. **48** Frank Spooner. **49** FLPA/Silvestri, TL; BCL/Dieter & Mary Plage, B; NHPA/David Woodfall, CR. **50** NHPA/Joe Blossom, TL; OSF/Lynn M. Stone, BR. **51** BCL/Charles & Sandra Hood, TL; OSF/Frances Furlong, BR. **52** FLPA/ E. & D. Hosking, BL; OSF/Kenneth Day, BR. **53** Siena Artworks Limited/Nigel Buckels. **54** BIOS/Dominique Halleux, R; BIOS/Jean Mayet, TL. **55** DRK Photo/Wayne Lynch, TL; OSF/Doug Allan, TC; OSF/Mike Birkhead, BR. **56** DRK Photo/Stephen J. Krasemann, T; DRK Photo/Wayne Lynch, BR. **57** OSF/Sinclair Stammers, TL; DRK Photo/Jeff Foott, C. **58** FLPA/S. Jonasson, T; Planet Earth Pictures/Margaret Welby, BL. **59** DRK Photo/John Gerlach, TR; FLPA/Roger Hosking, BL; NHPA/Alan Williams, BR. **60** NHPA/Henry Ausloos. **61** BIOS/Dominique Halleux, TL; Auscape International/D. Parer & E. Parer-Cook, B. **62** OSF/Kenneth Day. **63** OSF/Mark Hamblin, T; OSF/Colin Milkins, B. **64** OSF/Kim Westerskov. **65** Tom Stack & Associates/Milton Rand, T; OSF/G.I. Bernard, BR. **66** Siena Artworks Limited/Sharon McCausland. **68** Tom Stack & Associates/J. Lotter. **69** BCL/Jeff Foott, T; FLPA/D.P. Wilson, TR; OSF/G.I. Bernard, BL. **70** OSF/G.I. Bernard, TC; Biofotos/Heather Angel, TR; Planet Earth Pictures/Norbert Wu, BL. **71** Biofotos/Heather Angel, BL. **72** Biofotos/Heather Angel, T; FLPA/D.P. Wilson, B. **73** FLPA/D.P. Wilson, BL; NHPA/Leslie Jackman, TL; Biofotos/Heather Angel, BR. **74** DRK Photo/Norbert Wu, T; Auscape International/Becca Saunders, CL. **75** Biofotos/Heather Angel, TC; BCL/Charles & Sandra Hood, TR; NHPA/Roger Tidman, BL. **76** Tom Stack & Associates/Jeff Foott. **77** Planet Earth Pictures/Norbert Wu, TL; NHPA/B. Jones & M. Shimlock, CR; Auscape International/Jean-Paul Ferrero, B. **78** NHPA/B.A. Janes. **79** FLPA/M.J. Thomas, T; NHPA/G.I. Bernard, BL. **80** NHPA/Daniel Heuclin, T; NHPA/Anthony Bannister, BR. **81** FLPA/Michael Rose, TL; Planet Earth Pictures/Richard Coomber, TR; FLPA/R Tidman, C. **82** NHPA/James Carmichael Jr, TL; OSF/David Thompson, BR; NHPA/Roger Tidman, CR. **83** DRK Photo/Stephen J. Krasemann. **84** DRK Photo/Stephen G. Maka, TL; OSF/Peter Clarke, BL. **85** Biofotos/Heather Angel, TL; Siena Artworks Limited/Leslie Smith, BL; Planet Earth Pictures/David George, CR; Planet Earth Pictures/Mark Mattock, BR. **86** OSF/Roger Jackman, TL; OSF/Dr J.A.L. Cooke, TC; Planet Earth Pictures/Christian Petron, BR. **87** Planet Earth Pictures/Christian Petron, T; Planet Earth Pictures/Norbert Wu, BR. **88** Tom Stack & Associates/David Fleetham, T. **88-89** Auscape International/Ben Cropp, B. **90** DRK Photo/Johnny Johnson, T; Auscape International/D. Parer & E. Parer-Cook, CL; Auscape International/Jean-Paul Ferrero, BR. **91** OSF/Joaquín Gutiérrez Acha, TL; B. & C. Alexander, BR. **92** Planet Earth Pictures/John Bracegirdle. **93** OSF/Niall Benvie, TL; DRK Photo/J. Wengle, TR; OSF/Niall Benvie, BR. **94** OSF/David G. Fox, TL; FLPA/Chris Mattison, TR; Ardea/Ake Lindau, BL; Ardea/Su Gooders, BC. **95** NHPA/David Woodfall. **96** NHPA/L. Campbell, T; Planet Earth Pictures/Mark Mattock, C; Ardea/R. Gibbons, BL. **97** NHPA/David Woodfall, T; OSF/Deni Bown, BR. **98** FLPA/Tony Wharton, TL; OSF/G.I. Bernard, TR; NHPA/Laurie Campbell, BL. **99** Planet Earth Pictures/Thomas Broad. **100** FLPA/Roger Wilmshurst, TL; Planet Earth Pictures/Richard Coomber, BR. **101** OSF/R. Templeton, TL; **101** Planet Earth Pictures/Pete Oxford, BR. **102** Auscape International/D. Parer & E. Parer-Cook, T; NHPA/Anthony Bannister, BC; OSF/Michael Fogden, BR. **103** NHPA/Alberto Nardi. **104** OSF/Jack Dermid. **105** Auscape International/Jeff Foott, TL; NHPA/Kevin Schafer, TR; NHPA/Stephen Krasemann, BL. **106** OSF/Niall Benvie. **107** Auscape International/C. Gautier, TR; OSF/Kathie Atkinson, BL; NHPA/Jeff Goodman, BR. **108** OSF/Rodger Jackman, TL; Planet Earth Pictures/Mike Read, B; FLPA/Roger Wilmhurst, CR. **109** NHPA/A.N.T., TL; Planet Earth Pictures/John Lythgoe, BL; OSF/Norbert Wu, BR. **110** Planet Earth Pictures/John Lythgoe, TL; FLPA/Ian Rose, TR; FLPA/Ian Rose, BR. **111** OSF/David Boag, TL; OSF/Dr C.E. Jeffree, TC; OSF/S. Nagaraj, BR. **112** Auscape International/Jaime Plaza Van Roon, T; Planet Earth Pictures/Peter Scoones, CL; Ardea/Peter Steyn, BR. **113** OSF/Aldo Brando, TR; OSF/William Gray, TR Inset; OSF/Victoria Stone, BR. **114** Planet Earth Pictures/Armin Svoboda, BL. **115** Auscape International/D. Parer & E. Parer-Cook. **116** OSF/Doug Allan, TL; Ardea/Edwin Mickleburgh, TC; NHPA/Norbert Wu, BL. **117** OSF/Tui de Roy, BL; Ardea/ G. Robertson, TR. **118** B. & C. Alexander/H. Reinhard. **119** Ardea/Graham Robertson, TL; B. & C. Alexander, BC; OSF/Kim Westerskov, BR. **120** FLPA/C. Carvalho, TL; OSF/Doug Allan, BL. **121** Planet Earth Pictures/Jonathan Scott, TL; OSF/Doug Allan, TC; B. & C. Alexander, BR. **122** OSF/Doug Allan, TL; B. & C. Alexander, BL, BR. **123** B. & C. Alexander, TC, TR; OSF/Dan Guravich, BR. **124** OSF/Dan Guravich, TL; OSF/Norbert Rosing, B. **125** OSF/Stan Osolinski, TL; Tom Stack & Associates/Dave Fleetham, BR. **126** Auscape International/Tui de Roy. **127** Auscape International/Tui de Roy. **128** Auscape International/Tui de Roy, TL, TR, BR. **129** OSF/Tui de Roy, BL, TR. **130** Auscape International/François Gohier, TL; Planet Earth Pictures/Georgette Douwma, BL; OSF/Peter Parks, BR. **131** Planet Earth Pictures/Antony Joyce, TL; OSF/Margot Conte, BL; FLPA/D. Cavagnaro, BR. **132** Auscape International/François Gohier, TL; NHPA/David Middleton, CR; OSF/Maurice Tibbles, BR. **133** Auscape International/Ben Cropp, TL; Auscape International/Jean-Paul Ferrero, BR. **134** FLPA/J.C. Muñoz, TR; Tom Stack & Associates/Denise Tackett, BR. **135** Auscape International/Jan Aldenhoven, TL Inset; OSF/Michael Pitts, TL; Planet Earth Pictures/Brian Kenney, BR. **136** NHPA/Julie Meech. **137** Auscape International/Jean-Paul Ferrero. **138** NHPA/David Middleton, TL; FLPA/T. & P. Gardner, TC; NHPA/B. Jones & M. Shimlock, BL; Auscape International/Becca Saunders, CR. **139** Auscape International/Kev Deacon, TR; DRK Photo/Norbert Wu, BR. **140** OSF/David B. Fleetham, T; Tom Stack & Associates/Denise Tackett, CL. **141** FLPA/Susan Blanchet/Dembinsky, TL; OSF/Fred Bavendam, TR; OSF/Michael Pitts, BR. **142** Planet Earth Pictures/Bill Wood, TL; Planet Earth Pictures/Gary Bell, BL. **143** OSF/Peter Parks. **144** Planet Earth Pictures/Georgette Douwma, TL; Auscape International/Kev Deacon, B; OSF/Jamie Oliver, CR. **145** NHPA/Scott Johnson, TL; NHPA/Gerard Lacz, BR. **146** Planet Earth Pictures/Peter Scoones, B. **147** Planet Earth Pictures/Peter Scoones. **148** FLPA/Earthviews, TL; Auscape International/Jean-Paul Ferrero, TR; Auscape International/Tui de Roy, BR. **149** Auscape International/D. Parer & E. Parer-Cook, TL; NHPA/Rod Planck, BR. **150** OSF/Ben Osborne, TL; B. & C. Alexander/David Rootes, BR. **151** OSF/Daniel J. Cox. **152** OSF/Ben Osborne, T; OSF/Doug Allan, C. **153** OSF/Tui de Roy. **154** Auscape International/Tui de Roy, TL; Auscape International/D. Parer & E. Parer-Cook, TC; OSF/Kjell Sandved, BR; **155** Auscape International/Tui de Roy.

Front Cover: DRK Photo/Michael Fogden, background; BBC Natural History Unit/Tim Edwards, inset.